广州市对外宣传丛书

广府和味
Delicacies of Guangzhou

广州市人民政府新闻办公室 主编

广州新华出版发行集团
广州出版社

图书在版编目（CIP）数据

广府和味 / 广州市人民政府新闻办公室主编 . — 广州：广州出版社，2019.5
（广州市对外宣传丛书）
ISBN 978-7-5462-2907-2

Ⅰ.①广… Ⅱ.①广… Ⅲ.①饮食— 文化— 广州 Ⅳ.① TS971.202.651

中国版本图书馆CIP数据核字（2019）第083972号

广府和味
GUANGFU HEWEI

主　　编	广州市人民政府新闻办公室
出版发行	广州出版社
	（地址：广州市天河区天润路87号广建大厦9楼、10楼
	邮政编码：510635　网址：www.gzcbs.com.cn）
责任编辑	刘宗贵　卢嘉茜　刘甲
责任校对	黄焕姗　龚莉莎　李珊　罗嘉婧
装帧设计	广州晖恒广告策划有限公司
印刷单位	广州市赢彩彩印有限公司
	（地址：广州市白云区嘉禾街鹤边鹤泰东路工业区C栋
	邮政编码：510440）
规　　格	889 mm×1194 mm　1 / 32
印　　张	6.75
字　　数	160 千
版　　次	2019年5月第1版
印　　次	2019年5月第1次
书　　号	ISBN 978-7-5462-2907-2
定　　价	58.00 元

如发现印装质量问题，影响阅读，请与承印厂联系调换。

前言

　　一城之魅力，在于和美与共，无论是城中的人与人之间，又或是人与城市之间。

　　广州人实现"和美"的方式，当数"和味"。这一"和味"，在粤语里有"好味道"的意思，更有"搭配得法""食得合时""吃得享受"等含义。"和味"一词内涵之丰富，恐怕并非语言能全然表达，若想真正理解这个词，最好还是来广州"食游"体验一番。

　　有意无意间，广府文化中的"和味"与中华文化中的"天时、地利、人和"蕴含着某种微妙的联系。要想吃得"和味"，就必须顺应"天时"吃个新鲜，还得找到风味正宗的"地利"，最后还得有合适的餐伴，在饭桌上与宽容友善的广州人结为好友，方得一次味蕾和心灵相通的"和味"享受。

　　全世界不同的美食文化在广州相遇，存异求同——异的是可供相互借鉴的文化特质，同的是对美好生活的共同追求。而广府人自身则以其特有的淡定和生猛应对了这一切：正因为淡定，所以能包容多彩的外来食材；正因为生猛，所以能积极与"他者"的烹饪文化交融互通。淡定和生猛共同构成了这个城市的人心——淡定得像茶楼里的那杯茶，温热不烫嘴；生猛得像餐桌上丰富的菜肴，多彩而又相通。

　　特别感谢曾经为这本书贡献宝贵智慧的陈勋老前辈，以及其他对广州美食文化颇有研究的各位老师与前辈，如杨浩益、区又生、庄臣、冼悦华、利永周、蓝小青、饶原生、郭婉华、费勇、陈厚彬等。

　　最后，也是最重要的，感谢广州这个城市，感谢这个时代。正是广州人2000多年来对美好和美味生活的不懈追求，才让"食在广州"不断绽放新光彩。倘若这本书能为这块金字招牌增一点点光，实乃我们的幸运。

编者

2019年5月

Foreword

    The charm of a city lies in harmony, both among people in the city and between people and the city itself.

    Natives of Guangzhou seek harmony in "delicacies", which in Cantonese means the superiority not only in flavor but also in the combination of different ingredients, the timing and the way food is enjoyed. In a word, its rich meaning defies description. For a true understanding of the word, you'd better come to Guangzhou for a taste while travelling.

    Intentionally or not, there is a subtle connection between the "delicacies" in Guangzhou and the "right time, right place and right people" in the Chinese culture. To find "delicacies", one has to obtain fresh and authentically flavored ingredients from the "right place" at the "right time", and then share the food with the right companions and make friends with the warm-hearted Guangzhou people at dinner tables. Only in this way can you enjoy the "delicacies" that bring joy to both your stomach and your heart.

    Different food cultures meet in Guangzhou to seek common ground, different in their characteristics that can be learnt from each other and common in their shared goal of a better life. Natives of Guangzhou cope with all of these with their characteristic calm and vigor. It is because of such calm that they can integrate colorful ingredients from other places, and it is because of such vigor that they can actively engage and combine with other food cultures. Calm and vigor combine to represent the people in the city: they are as calm as the cup of tea in a teahouse, warm but not too hot; and as vigorous as the rich dishes on the table, colorful and inclusive.

    My special thanks go to Mr.Chen Xun, who has contributed a lot of wisdom to the book, and other experts on Guangzhou food culture including Yang Haoyi, Ou Yousheng, Zhuang Chen, Xian Yuehua, Li Yongzhou, Lan Xiaoqing, Rao Yuansheng, Guo Wanhua, Fei Yong and Chen Houbin.

    Last but not least, we feel grateful to Guangzhou in this era. It is the natives of Guangzhou that have continuously pursued happiness with delicacies for more than two thousand years, which makes Eating in Guangzhou become a bright cultural symbol of Guangzhou. It will be our honor if this book helps this cultural symbol glitter.

Author
May 2019

目录 Contents

第一章 | Chapter I 时天 Right Time

第一节 | 食得新鲜·春 　　　　　　　　　　003
Section I | Eating Freshly ◆ Spring

嫣红礼云子，只为春来艳 　　　　　　　　003
Bright Red Liyunzi Only Shows Up in Spring

春来鱼鲜跃 　　　　　　　　　　　　　　011
Eating Fish in Spring

第二节 | 食得新鲜·夏 　　　　　　　　　　017
Section II | Eating Freshly ◆ Summer

鲥鱼，时也；多刺，无惧也 　　　　　　　017
Reeves Shads Should Be Eaten at the Right Time;
No Fear For Its Bones

蟹中杨贵妃，最美黄油肥 　　　　　　　　025
"Imperial Concubine Yang" in the Crabs, the Richest Butter Crabs

荔枝，不只是佳果 　　　　　　　　　　　031
Lychee, More Than a Good Fruit

第三节 | 食得新鲜·秋 　　　　　　　　　　043
Section III | Eating Freshly ◆ Autumn

秋风乍起时，只追禾虫肥 　　　　　　　　043
When the Autumn Winds Blow, Rice Worms
Are at Their Best

第四节 | 食得新鲜·冬 　　　　　　　　　　049
Section IV | Eating Freshly ◆ Winter

"虽迟但到"菜心味 　　　　　　　　　　　049
Late but Right Choy Sum (Chinese Flowering Cabbage)

炭火炉带来的冬日暖意 　　　　　　　　　057
Winter Warmth by the Charcoal Stove

第五节 | 食得健康 　　　　　　　　　　　　063
Section V | Eating Healthily

饭菜如舟，靓汤如水 　　　　　　　　　　063
Rice is Like a Boat While Good Soup is Like Water

凉茶的甘苦之道 　　　　　　　　　　　　075
The Taste of Herbal Tea

"治愈系"的"广州甜" 　　　　　　　　　085
Syrup for Cure

第二章 | Chapter II
地利 Right Place

第一节 | 米的幻化 — 095
Section I | Transformations of Rice

肠粉"奏鸣曲" — 095
The Rice Roll "Sonata"

非遗一箸沙河粉 — 103
The Intangible Cultural Heritage - Shahe Rice Noodles

"濑"出米的"味"力 — 109
"Flowing" Tastes

淡薄粥中滋味长 — 117
Long-lasting Flavor in Insipidity of Porridge

第二节 | 小吃不小 — 125
Section II | More than just Snacks

云吞面，是面还是云吞？ — 125
Wonton Noodles? Noodles or Wonton?

广府小饼这一家 — 137
Cantonese Pastry

第三节 | 风味荟萃 — 147
Section III | Combining Various Flavors

"有鸡味" — 147
"Chicken Flavor"

腊味：来自北回归线的馈赠 — 151
Preserved Meat: a Gift from the Tropic of Cancer

吃出那"啫啫"的声音了吗？ — 157
Have You Tasted the *Ger Ger* Sound?

在广州，包容万菜 — 163
Inclusive Cantonese Dishes Featuring the West and the East

第三章 | Chapter III
人和 Right People

第一节 | 茶楼百态 — 177
Section I | Life in the Teahouse

得闲饮茶 — 177
Enjoy Drinking Tea at Leisure

一盅两件：粤食"万花筒" — 185
One Pot and Two Pieces: Cantonese Food Kaleidoscope

第二节 | 食有亲朋 — 191
Section II | Sharing with Families and Friends

流光溢彩广府家宴 — 191
The Colorful Cantonese Family Feasts

消夜是广州人的仲夏夜之梦 — 201
Midnight Snacks: A Midsummer Night's Dream for Guangzhou Natives

第一章 | Chapter I
Right Time

天时

广州人讲究"不时不食",一年四季中,无论是时令食材的选择、养生靓汤的烹制,还是健康凉茶的熬煮等,都体现了他们与自然和谐共生的独特生活理念及文化观念。

The natives of Guangzhou attach such great importance to the concept of "choosing right food at the right time" that their choices of seasonal ingredients, their cooking of soups and their brewing of herbal teas for health all reflect their unique attitude on life and culture of harmonious coexistence between people and nature.

■ 油盐清蒸礼云子（陈枫摄影）

第一节 | 食得新鲜 ◆ 春
Section I | Eating Freshly ◆ Spring

嫣红礼云子，只为春来艳

Bright Red Liyunzi
Only Shows Up in Spring

礼云子其实就是蟛蜞之卵，产于南番顺等地的水田之间，出没于春末之时，是广东独特的鲜物。礼云子呈红棕色，口感幼滑细腻，鲜味极浓。灿若云霞的嫣红色，齿颊充盈的浓鲜味，一年只等这一回的相逢。

The Liyunzi is actually the eggs of amphibious crabs, which are produced in the paddy fields of Nanfanshun (currently part of Guangzhou and Foshan) at the end of spring. It is a unique delicacy for the Cantonese, with a reddish brown color, smooth and delicate taste, and rich flavor.

一代美食大家江献珠先生偶读《广州西关古仔》发现了"礼云子"一名的来源。广州话"来"与"礼"谐声，蟛蜞卵煮熟后犹如天边的云彩，因此人们戏称蟛蜞卵为"来云子"，尔后文人厨者将它写上菜牌，字面上书为"礼云子"。

■ 礼云子蒸豆腐（陈枫摄影）

Ms. Jiang Xianzhu, a gastronome, accidentally found the origin of the name "Liyunzi" when she read *Stories of Xiguan, Guangzhou*. The cooked eggs of amphibious crabs are just like bright clouds in the sky, so people nicknamed the food "Laiyunzi"(literally meaning "comes the cloud"), which was later written as "Liyunzi" ("Li" sounds like "Lai" in Cantonese) on menus by literary cooks.

2001年，香港美食家唯灵先生于镛记餐厅举行的主题为"乡亲带来新造礼云子"的宴席上，聊起礼云子名字的来源。同席的香港中文大学教授陈胜长先生称古人见面拱手为礼，拱手之状如同小蟛蜞横行。对此，番禺美食家屈九先生后来补充道：春天蟛蜞有卵，为保护卵，蟛蜞偶尔直行，两只前螯合抱，

天时 | 第一节 | 食得新鲜 ◆ 春
Right Time | Section I | Eating Freshly ◆ Spring

一步一叩首，如同古人行礼作揖。因此取孔子之名言"礼云"，蟛蜞有了"礼云"之名，其子自然就叫"礼云子"。

In 2001 at the themed banquet of Yong Kee Restaurant, Hong Kong Gourmet Mr. Mark Yiu Tong mentioned the origin of the name Liyunzi. Prof. Chen Shengchang from the Chinese University of Hong Kong, who was at the dinner, explained that in ancient China, people greeted each other with their hands folded in front, much like sideways crabs. To this, Gourmet Qu Jiu from Panyu later added that in spring, amphibious crabs bear eggs. In order to protect them, the mother crabs sometimes walk straight with their two pincers folded in front, just like people greeting. Therefore amphibious crabs were named Liyun from Confucian doctrines, and their eggs were accordingly named Liyunzi.

美食家杨浩益在沙面长大，他至今仍然记得儿时放学归来到涌边田里捉蟛蜞的场景，一抓一大把，实为童年一大乐趣。

今时今日，礼云子已经成为餐桌上的珍稀食材，珍稀只因"季节甚短，稍纵即逝"（江献珠先生语）。腰记餐厅老板欢姐说，10斤（1斤=500克）的蟛蜞才得三四两（1两=50克）礼云子。广州供应正宗礼云子的餐厅屈指可数，过去珠三角有店家以蟛蜞膏充礼云子，不知者懵懂就食。然而二者非同一物，一是膏，一是卵。

Gourmet Yang Haoyi, who grew up in Shamian of Guangzhou, still remembers his happy childhood when he caught amphibious crabs in the field or by the riverside.

Nowadays, Liyunzi has become a rare and precious food on the table, because of "the fleeting season" (by Jiang Xianzhu). According to the boss of Yaoji Restaurant, only 150 to 200 grams of Liyunzi can be obtained from 5 kilos of amphibious crabs. Only very few restaurants in Guangzhou offer authentic Liyunzi. In the past some restaurants in the Pearl River Delta passed crab roe off as crab eggs.

春季为蟛蜞交配之时，此时蟛蜞满腹卵子，这些卵是极其幼细的、如沙状，比虾子还细。农民从田里捉到蟛蜞后取卵洗净，放于瓦盅内，用盐保鲜。那时还没有冰箱，礼云子不能久存。这一季节性极强的食材，常在清明前一个月露面，因此又有"春水礼云子"之说。

In spring, amphibious crabs mate and bear eggs, which are even finer than shrimp eggs. Peasants kept cleaned amphibious crabs' eggs in pottery and preserved them with salt. Without refrigerators, the eggs could not be kept for long, so this seasonal food was only available one month before the Qingming Festival, hence its another name for it: "Spring Liyunzi".

天时 | 第一节 | 食得新鲜 ◆ 春
Right Time | Section I | Eating Freshly ◆ Spring

晚清时期，美食家江孔殷的家中常用礼云子做馅，将它或包裹于春卷之中，制成礼云子春卷，或制成礼云子薄饼，或制成礼云子粉粿，只供江孔殷一人食用。礼云子也可用来捞面或者炒蛋。江献珠先生在她的书中提及，祖父江孔殷令家中厨子亚勋用礼云子炒蛋，三次方"收货"。第一次批评蛋未熟，油过多；第二次批评过熟；第三次方合格。至于那些炒坏了的礼云子，则用来炒饭。那鲜香美味的炒饭让江献珠先生毕生难忘。

In the late Qing Dynasty, at Gourmet Jiang Kongyan's home, Liyunzi was often made into spring rolls, pancakes or glutinous cakes, exclusively served to Jiang himself. Liyunzi could also be used in *lo mein* or fried eggs. Ms. Jiang Xianzhu mentions in her book that his grandfather Jiang Kongyan did not accept the food until the cook made the fried eggs for the third time.  The first time it was criticized as too raw and too oily; the second time overdone and only the third time was up to standard. The rejected fried eggs with Liyunzi were used to make stir-fried rice, whose fresh flavor is remembered all her life.

今天，我们有幸能品尝到礼云子之味。广州番禺的腰记饭店有多年烹饪礼云子的经验，老板欢姐甚至设计了十道菜肴，做成礼云子宴。以礼云子现在的市价，窃以为最奢侈的吃法是"空口吃礼云子"。油盐清蒸礼云子，没有任何配菜，吃一勺相当于吃掉了几十元！当然，一般是配着白饭吃的。柚皮礼云子与柚皮虾子制法相似，取沙田柚的柚皮，蒸熟后挤干水，焖软之后加入礼云子煮熟。礼云子的鲜丝丝渗入柚皮中，此时柚皮柔软而不干涩。通常粤菜师傅处理柚皮时都要削去黄色外皮，但欢姐弃用此法却能收到成效。礼云子炒通菜以及礼云子煎蛋，则令礼云子接受大火的洗礼，愈加逼出其鲜味。礼云子豆腐、礼云子蒸蛋、礼云子蒸鱼、礼云子蒸花腩肉和礼云子蒸腊味的做法简单，将礼云子铺陈其上蒸制即可。

Today, we are lucky enough to taste Liyunzi. Yaoji Restaurant in Panyu has years of experience in cooking Liyunzi. Its boss has even designed a banquet of 10 dishes of Liyunzi. Considering the market price, I think the most luxurious way to enjoy Liyunzi is to eat it alone, that is

■ 柚皮烩礼云子（陈枫摄影）

天时 | 第一节 | 食得新鲜 ◆ 春
Right Time | Section I | Eating Freshly ◆ Spring

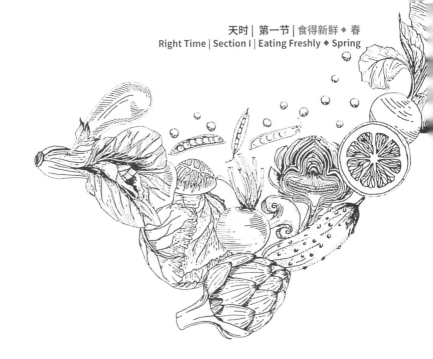

steaming Liyunzi with salt and oil without other ingredients. One spoonful is worth several dozen yuan. Of course, it is usually served with rice. "Liyunzi with pomelo peel" is made in a similar way as "shrimp spawn with pomelo peel". After being steamed, strained and stewed, pomelo peel is cooked with Liyunzi. The freshness of Liyunzi is infiltrated into the peel, which is soft. Usually the outer layer of the pomelo peel needs to be removed first in Cantonese cuisine, but Yaoji Restaurant achieves the same effect without following the tradition. Stir-fried water spinach with Liyunzi and Liyunzi omelets taste best because high heat is used. Liyunzi tofu, Liyunzi steamed eggs, Liyunzi steamed fish, Liyunzi streaky pork and steamed preserved meat with Liyunzi are all easy to make, just by spreading it over other ingredients before steaming.

春来鱼鲜跃
Eating Fish in Spring

阳春三月，春光无限好，最是鱼儿肥美时。鲜品鱼类，生于水而长乎水，一切活动皆与水息息相关。好鱼必从好水出。广州番禺、南沙地处珠江入海口，占尽地利。在咸淡水交界的入海口产出的鱼，既得河鲜的生猛新鲜，又得海鲜的肥美鲜甜。和大名鼎鼎的黄沙水产市场相比，地处郊区的南沙十八涌、十九涌也丝毫不逊色。这里有许多野生限量版的水产，而这些"生鲜猛料"多为小船、小艇作业所得，量极少，"鲜货"一起水便被一抢而空。若想拿到渔民手中的水产，除非打"熟人牌"。因此一到周末，南沙十八涌、十九涌、番禺海鸥岛等地便吸引了众多广州食客前往，只为一口生猛河鲜。

March is the best time to eat fish. Good fish cannot live without good water. Panyu and Nansha, located at the mouth of the Pearl River, enjoy the benefit of both fresh water and sea water where fish are robust, fresh and delicious. Compared with the renowned Huangsha aquatic product markets, 18 and 19 Chung Nansha in the suburbs is by no means inferior. In this market, there are many wild aquatic products in limited supply, mostly obtained from small boats. They are so popular that buyers have to solicit help from acquaintances. Therefore, many foodies in Guangzhou are attracted to 18 and 19 Chung Nansha and Panyu Seagull Island on weekends for the delicious river fish.

　　刚刚经历了秋冬贴肥膘，人们都在嚷嚷着减肥，但河鲜却大大不同，春天到了也要继续"发福"，哪怕"肚子"胖上两圈也在所不惜。这里说的河鲜便是曹虾。被称为"虾中之王"的曹虾盛产于南沙十八涌的咸淡水交界处，4月中旬是其最肥美的时期，只只肥得足有拇指大。曹虾虾壳细而薄，不需要像其他虾那样剥壳吃，整只虾连壳放入口中咀嚼，最后连壳一起吞咽下肚即可，省心省力。因此，曹虾在广州人心中评价极高，又有"千虾万虾不如曹虾"的说法。

　　In spring, people may be obsessed with losing weight after gaining it in autumn and winter, but Cao Shrimp is different. It is time for it to get fatter. This "king of shrimps" lives at the junction of fresh water and sea water at 18 Chung Nansha . Mid-April is its mature time. The shrimp can be as big as a thumb. Its shell is so fine and thin that the whole shrimp can be swallowed without shelling, which is convenient. Therefore, Cao Shrimp enjoys high popularity among the natives of Guangzhou, who comment that "East and west, Cao Shrimp is the best."

天时 | 第一节 | 食得新鲜 ◆ 春
Right Time | Section I | Eating Freshly ◆ Spring

　　加热后的曹虾，虾壳透明并带有淡淡橙红色，可见嫩白的虾肉包裹着橙红或微褐色的虾子。雌的曹虾比雄的好吃，因为其丰润饱满有虾子。而虾子呈红色代表这虾刚刚成熟，鲜嫩无比；虾子呈褐色则代表虾快排卵了，口感软糯。而曹虾最传统的做法，便是白灼。起锅放少许油盐，拍一两粒蒜，放少许水。待水一滚，立即下虾，最后下少许胡椒粒提味。

　　After being heated, the shell is transparent with a bit of light orange, and beneath its white meat is orange or brownish spawn. Female Cao Shrimps taste better than male ones because of the spawn. Orange spawns indicates that the shrimps are at their right time, so they are extremely fresh and tender; brown spawn indicates that they are over mature, so their spawn tastes sticky. The best way of cooking Cao Shrimps is to boil them. Just add some oil and salt, one or two cloves of garlic and a little water to the pot. As soon as the water boils, add the shrimps. Finally add some pepper while seasoning.

春初，三月回暖，此时凤尾鱼悄悄出没。生长于南沙咸淡水交界处的凤尾鱼尾部呈剪刀状，像凤凰的尾巴。它是一种洄游性小型鱼类，通常一条才一两多，若是有鱼卵，则肥厚许多。凤尾鱼的做法多，最传统的莫过于蒜头豉汁蒸，鱼肉入口即化。若用作煎焗，香而不腻。在南沙，凤尾鱼卵更是一道名菜，被很多美食老饕称为"堪比法国鹅肝的美味"。风干后的凤尾鱼卵金黄透明，甘香可口，佐以姜蒜等清蒸，味道极鲜。

In March, the weather becomes warmer, so anchovies begin to appear. The anchovy at the junction of fresh water and sea water at Nansha has a tail like that of a phoenix. It is a small, migratory fish. It weighs just over 50 grams. If it bears eggs, it is much heavier. There are many ways of cooking the anchovy. The most traditional way is to steam it with garlic and fermented black beans, making it so tender to taste. If it is fried or simmered, it smells nice but is not greasy. In Nansha, anchovy eggs is a famous delicacy, which has been praised by many foodies as "a delicacy comparable to French foie gras". Air-dried anchovy eggs are golden and transparent. When steamed with shredded ginger and garlic, anchovy eggs are extremely fresh and delicious.

■ 榄角蒸鳊鱼

俗话说：春鳊秋鲤夏三黎。阳春三月，鳊鱼当季。塘鱼色泽较深，野生鳊鱼则稍显色淡。骨丝多的鱼通常比较鲜，鳊鱼正属此类。在番禺，鳊鱼的经典做法是用榄角蒸，榄角的馥郁甜香在炉火的滋养下渗透入鱼鲜。

As the saying goes, the best fish is the bream in spring, carp in autumn, and Sanli Fish in summer. March is the mature time of the bream. Pond breams are darker while wild ones are brighter. Usually fish with more bones are more delicious. Breams are just such fish. In Panyu, the classic way of cooking the fish is to steam it with preserved olives, whose aroma and sweetness penetrate into the fish by heating.

在番禺，还有一种通体无鳞的鱼儿颇受欢迎，如泥鳅一样拥有让人羡慕的光滑肌肤，一条大约一两重，它便是奶鱼。其脊骨像蛇骨一样硬朗，但小骨较多，因此奶鱼不太适合蒸、焗，更适合煮粥或者做汤，那鲜美的滋味真叫人忍不住多吃两碗。

In Panyu, a kind of little scaleless fish is very popular. It has smooth skin like a loach. It weighs just about 50 grams. It is called milk fish. It has a backbone as hard as that of snakes and many small bones, so instead of steaming or simmering, it is more suitable for cooking porridge or soup, whose delicious taste invites you to eat more.

第二节 | 食得新鲜 ◆ 夏
Section II | Eating Freshly ◆ Summer

鲥鱼，时也；
多刺，无惧也

Reeves Shads Should Be Eaten at the Right Time;
No Fear for Its Bones

人生有三大遗憾：鲥鱼多刺，海棠无香，《红楼》未完。

——张爱玲《红楼梦魇》

"There are three regrets in life: the delicious reeves shad has too many bones; the beautiful Malus spectabilis is not fragrant; and the great *Dream of the Red Chamber* is not finished."

——*Nightmare in the Red Chamber*, Eileen Chang

1977年，张爱玲的学术考据之作《红楼梦魇》出版，这是张爱玲1966年前后开始的中国古典小说研究的成果。其时，作家身在美国，也许是异乡单调乏味的食物让她对家乡的鲥鱼之鲜多了几分眷恋，于是就在书中写下了"人生有三大遗憾：鲥鱼多刺，海棠无香，《红楼》未完"。其中提到的鲥鱼，主要产自华东和华南一带，正是张爱玲在国内的常住之地——上海、广州和中国香港。在广东，鲥鱼也被称为"三黎鱼"。

Published in 1977, *Nightmare in the Red Chamber* is the fruit of Eileen Chang's academic research starting around 1966 on Chinese classic novels. She lived in the United States at that time. It might be the bland food in the foreign land that made her miss the flavor of the reeves shads in her hometown. She thus wrote, "There are three regrets in life: the delicious reeves shad has too many bones; the beautiful Malus spectabilis is not fragrant; and the great *Dream of the Red Chamber* is not finished." The reeves shad mentioned mainly comes from East and South China, Chang's long-term residence in China, namely, Shanghai, Guangzhou and Hong Kong. The fish is also called "Sanli Fish" in Guangdong Province.

吃鲥鱼，最重天时合宜。因为鲥鱼平素生活在大海之中，只有每年4—7月会溯河而行、洄游生殖，产卵后在9—10月重新回到大海。鲥鱼初入江时脂肪特别肥厚，肉味也最为鲜美。而在产卵回海后，肉质便不再肥美，而且难以捕获。《本草纲目》云：鲥鱼"初夏时有，余月则无"，民间也有"春鳊秋鲤夏三黎"的说法。

It is vital to eat reeves shads at the right time because they usually live in the ocean. They will go upstream for breeding from April to July every year and return to the ocean in September and October after spawning. The reeves shads are particularly fat and delicious when they first go upstream to the rivers. But after they spawn and go back to the ocean, they are no longer fat and are difficult to catch. *Compendium of Materia Medica* says that reeves shads appear in early summer and disappear in the rest of the year. There is also a folk saying of "eating breams in spring, carps in autumn and Sanli Fish in summer".

顶好的食材，还要有好的厨师主烹，方能不暴殄天物。嘴刁的张爱玲有没有吃过粤菜师傅烹制的鲥鱼，大概已难以考证。不过,比起广州的食客们,张爱玲对鲥鱼的喜爱和研究就成了"小儿科"了。

Top ingredients need top cooks to present the best dishes. It is hard to prove whether picky Eileen Chang has ever enjoyed reeves shads cooked by Cantonese chefs. However, compared with the foodies in Guangzhou, her love and study of reeves shads would be not worth mentioning.

广州人在吃的方面崇尚天然，这在鲥鱼的烹制上表现得淋漓尽致。除了要抓住春末夏初的短短数周捕捞外，岭南的渔民们还会在刚打捞上来的鲥鱼身上涂抹盐巴保鲜，但决不让其于阳光下曝晒，此举是为了保持鲥鱼本身的鱼汁。聪明的粤菜师傅还发现鲥鱼的鱼鳞大有用处，如同天然的保鲜剂，将肥美的鱼油紧锁其中。因此烹鱼时，识货的粤菜师傅会保留鲥鱼鱼鳞。于是，天赐良时、天然鱼汁、肥美鱼油，便组成了一曲销魂的三黎鱼小调。

Natives of Guangzhou follow the laws of nature in eating, which is especially true when they eat reeves shads. Apart from taking advantage of the first few weeks of summer, the fishermen living in Lingnan spread salt over the fish to keep it fresh once it is caught, but will never expose the fish to the sun so that the juice of the fish can be preserved. Clever Cantonese chefs also think that the scales of the fish are very useful, which are just like a natural preservative film that can perfectly preserve the delicious fish oil. Therefore, the right time, natural fish sauce and rich fish oil constitute a wonderful ditty of "Sanli Fish".

广州的资深食客们心心念念一整年，怀着"错过等一年"的热情，以及无惧多刺多骨的大无畏精神，无非为了那几个星期的鲜味享受。甚至，广州人根本不会像大才女张爱玲那样对"多刺"这件事耿耿于怀，因为在他们看来，就是"多刺的鱼才鲜美"，"人家天生就是这样的"。这一观念不仅体现在对鲥鱼的态度上，广府人对鲫鱼、鲮鱼等河鲜也是乐见其多刺的。张爱玲的嗟叹只是文人多愁善感而已，"喜闻乐吃"的食客大抵不会在意。

The senior foodies in Guangzhou will constantly think of reeves shads for a whole year with the fear of missing the time and having to wait for another year. They are not afraid of the fine bones because they will be satisfied with the savory taste of the fish for a few weeks. Natives in Guangzhou would not be as worried about the bones as Eileen Chang did, because in their view, it is the fish with

many fish bones that is delicious and the fish is naturally born so. This attitude is also shown when they are viewing freshwater fishes like crucian and dace. They believe that the bones in the fishes are natural and are willing to accept them. As for the sigh of Chang, the Guangzhou natives who care more about eating will probably think it mere sentiment of literati.

近年来，由于环境变化，野生鲥鱼日渐稀少，已成保护动物。大多数热衷于与自然和谐共处的广州人，不愿意继续捕捞，野生既少，那就养殖吧，竟也渐成一似模似样的产业。

In recent years, due to environmental changes, the increasingly scarce wild reeves shads have become endangered animals. Most people in Guangzhou who are keen on living in harmony with nature are unwilling to continue fishing wild reeves shads. As a result, reeves shad aquaculture has become an emerging industry.

粤菜师傅对天然食材的顺应和利用还不止于此。若张爱玲活到今天,大概就可以在其"遗憾清单"中去掉"鲥鱼多刺"一项了。鲥鱼在现今粤菜做法中,由于鱼肉多经过盐巴腌制,肉质变得结实,骨肉自然分离,所以吃起来感觉鱼刺变少了,也就没有了张爱玲所恨的"鲥鱼多刺"了。甚至还有些粤菜师傅,竟能让同样以多刺著称的鲫鱼变成"啖啖肉"。

The adaptation and utilization of natural ingredients by Cantonese chefs does not stop there. If Eileen Chang lived to this day, she would probably remove "reeves shads with too many bones" from her list of regrets. In the current Cantonese cuisines, reeves shads are mostly salted, and the flesh is firm and naturally separated from the bones. Therefore, it tastes like having fewer bones. Some Cantonese chefs even creatively transform the crucian which also has many bones into boneless meat.

奇妙变化,尽在刀功。要完全把一根根细小的鲫鱼刺完全挑出来是不可能的,但粤菜师傅却另辟蹊径:首先精选1~1.5斤重的鲫鱼,这个重量的鲫鱼肉质最鲜嫩,之后以高强的刀功将鲫鱼中部的鱼肉与鱼骨分离,其余的细小鱼刺则用刀功"内力"震碎、打断。如是者,虽然还有鱼刺保留在鱼肉内,但食客品尝时却已没有丝毫"刺感"了。更有融会贯通者,将无骨鲫鱼酿入陈村粉中,每片陈村粉都沾上了鱼肉的鲜香,却又保持着它本身的鲜甜软滑,一啖入口,满嘴都是鱼肉和米粉的交融柔情。

The wonderful transformation starts from the chef's excellent cutting and slicing skills. In fact, it is impossible to pick out every tiny bone of the crucian, but a master chef has managed to find a creative solution. He first selects a crucian with the weight between 500 to 750 grams, the most tender in texture. Then, he separates the flesh from the bones in the middle of the crucian and shatters or breaks other bones with his excellent skills and strength. In this way, the customers will not feel the bones when they are eating even though the bones are still in the flesh. There is also a creative way of filling boneless crucian flesh into Chencun rice sheets. Each piece of Chencun rice sheets is simmered with the fish, but retains its own features of sweetness, softness and smoothness at the same time. Once the customers have a bite of the rice sheets, their mouths will be full of a fusion taste of fish and rice sheets.

如此奇技，大概唯粤菜师傅才有心施展一番。
Probably only the Cantonese chefs would be caring enough to display such an amazing technique.

在尊重鲗鱼多刺的前提下，广州人发挥极大创意，在养殖和烹饪技术上下功夫。也许粤菜师傅和食客们说不出什么"天人合一"的大道理，但他们能说："嗯，这条鱼现在这个时节最'啱食'（广府俗语，意为'适宜吃'），就这段时间才够鲜甜的嘛。"谁又敢说广府人对"天人合一"的理解又丝毫逊色于儒道两家的研究者呢？

Under the premise of respecting the natural laws of eating reeves shads at the right time, natives in Guangzhou have unleashed the creativity and made great efforts in developing aquaculture and cooking techniques. Perhaps Cantonese chefs and the Guangzhou natives are unable to say great words like "man is an integral part of nature". But as long as they say something like "well, this fish is right to eat in this season and it is the most delicious during this time", no one can deny that they clearly understand "man being an integral part of nature" just as the Confucian and Taoist scholars do.

蟹中杨贵妃，最美黄油肥

"Imperial Concubine Yang" in the Crabs.
The Richest Butter Crabs.

在《鹿鼎记》中，江湖中人个个口称"平生不识陈近南，纵称英雄也枉然"，大有陈近南一出，便等于正式认证了好汉们的江湖身份一样。若是放在蟹界，这个蟹国状元必定是黄油蟹。

In *The Deer and the Cauldron*, every wandering person claims that "one can never claim to be a hero before he knows Chen Jinnan". It seems that Mr. Chen serves as the key to identify those heroes. Supposing it is in the crab world, the ultimate identification of NO.1 crab must be the butter crab.

为何一年到头，由奄仔蟹、重皮蟹、水蟹、六月黄到大闸蟹，偏偏就是黄油蟹担大梁？理由很简单：一是当造期短且数量稀少，就算是在水产丰富的广州，每日现身的黄油蟹都不过百只。二是它食味靓绝，肉香膏滑，仅以膏论，可说是无可匹敌的滑溜。这两点，就可奠定黄油蟹在蟹界的状元地位。

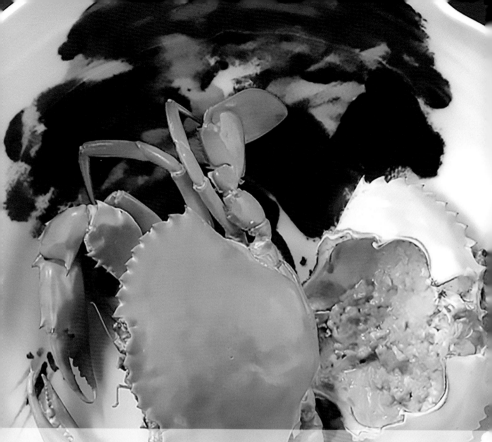

Why can butter crabs stand out as the leader all the year round among all the crabs from Yanzai crab, two-shelled crab, water crab, yellow crab of June immature hairy crab to hairy crab? Two simple reasons are at play for that. One is for its short season and scarce quantity. Even in Guangzhou, which abounds in aquatic products, there are no more than 100 butter crabs appearing in the market every day. The other is for its delicious taste, good flavor and soft crab roe. For crab roe only, the butter crab is unrivaled in smoothness. These two advantages help explain the position of the butter crab as "No.1" in the crab world.

■ 清蒸黄油蟹（曾繁莹摄影）

暑热时节，是食用黄油蟹的绝佳期。黄油蟹是真正的"百里挑一"，一百只成熟膏蟹中最后仅有一只成功转化成黄油蟹。凭着油膏甘香、肉质鲜嫩的特点，黄油蟹享誉南国，被称为"蟹中杨贵妃"，它是广东人甚为骄傲的自然馈赠。众所周知，黄油蟹需要充足的日晒，它不会被晒成古铜色，但是会出落得膏脂肥腴。在太阳暴晒下，高温导致壳中的蟹膏化成油状遍布全身，整只蟹从螯到脚端，甚至是关节处都布满黄油。浑身布满黄油的黄油蟹呈金黄色，沐浴阳光，绽放光芒，隐藏着甘香醇厚。黄油蟹以野生为上品，每逢蟹季，在广州南沙十八涌、十九涌一带，多少广州人慕名而来，只为那一只肥美甘香的野生黄油蟹。

　　The hot summer season is the best time for butter crabs. They are really "one in a hundred"—only one transforms successfully among 100 mature green crabs. Sweet roe and tender meat earn the butter crab a good reputation as "Imperial Concubine Yang" in southern China, which is also an endowment of nature, of which the Cantonese are proud. It is well-known that plenty of exposure to the sun is necessary for the butter crab. With enough sunshine, the butter crab will grow to be tender instead of becoming bronze by the sun. In the heat of the sun, the high temperature makes the crab roe in the shell oily all over the body, and the whole crab is covered with butter from pincers to toes, even joints. Bathed in sunshine, the butter crab turns golden yellow, shining with its sweet smell hidden inside. Wild ones are ranked as the top-grade. In every season for crabs, many Guangzhou natives come to 18 and 19 Chung of Nansha, Guangzhou, to enjoy that sweet wild butter crab.

若说黄油蟹是蟹中杨贵妃,那奄仔蟹就是养在深闺的窈窕少女。奄仔蟹是青蟹中的处女蟹,胜在肉质滑嫩,清丽脱俗,好比豆蔻年华的女子,嫩得能掐出水来。每年3—7月最当造,分青、白、黄、黑4级,以黑奄为最佳。盛夏时节,奄仔蟹多在台山、深圳和广州等地出现,南沙十八涌、十九涌也是其产地。它蟹膏绵软细滑,犹如流沙奶黄。有食家赞叹奄仔蟹为平民版黄油蟹,奄仔蟹与价值不菲的黄油蟹相比,每只最多100元出头。此外,奄仔蟹有未死先臭的说法,即有些不新鲜的奄仔蟹,表面看起来依然是鲜活的,但闻起来却有股臭味,这就表明该蟹的身体内某个器官已经坏死了。

■ 蒸奄仔蟹(王维宣摄影)

If butter crab is "Imperial Concubine Yang" in the crabs world, then Yanzai crab is like a fair lady kept in purdah. Yanzai crab is the maiden crab in green crabs with tender meat as its advantage, like a blooming girl. March to July every year is the mature season of the crabs, which are classified into four grades-green, white, yellow and black, among which black ones are the most delicious. Hot summers see Yanzai crabs appearing in districts including Taishan, Shenzhen and Guangzhou as well as 18 and 19 Chung of Nansha. Its crab roe is as soft and slippery as custard. Yanzai crab is praised by the diners as a "civilian" version of butter crabs. Compared with pricey butter crabs, Yanzai crab is no more than 100 yuan each. In addition, there is a saying that Yanzai crabs stink before they die, that is, some stale Yanzai crabs still look fresh on the surface, but they smell stinky, which indicates that some organs in the crab's body have failed.

"虾兵蟹将"多有坚硬的外壳，让部分牙口不好的朋友只好远观。然而，还有一种蟹，蟹壳软软的，连老人家都能轻易咬烂，它便是软皮蟹，是重壳蟹蜕壳之后的结果。当两层壳的重皮蟹为继续长大而蜕壳后，旧壳会被它们自己吃掉，新壳会在2~3天内软如纸张。此时蟹最脆弱，甚至你把手指送到螯下，它们都无力夹你。因此，这个时候它们往往把自己藏得很隐蔽，出现的概率也几乎为零。软皮蟹的蟹膏较滑溜，如刚蒸好的鸡蛋羹，略带微妙的酸，更凸显蟹肉的鲜甜顺滑。

Hard shells of shrimps and crabs have kept some eaters with bad teeth away. However, there is a species of crabs with such soft shells that it can be easily bited by the elderly. This is the soft-shelled crab, which has moulted its outer shell. When two-shelled crabs moult to continue growing, the old shell will be eaten by themselves, and the new shell will be as soft as paper for two to three days. At this time crabs are the most fragile so that they don't have enough strength to clamp you even if you put your fingers under their pincers. As a result, they often hide themselves and are rarely seen at this time. The crab roe of them is as smooth as steamed egg custard, slightly acidic, showing the fresh and sweet taste of crab meat.

在番禺莲花山码头一带，渔民有时还会捕获一些海蟹，例如大红蟹，块头大味道鲜。约上三两知己到莲花山码头或者海鸥岛挑选靓蟹、品蟹，然后对着海天一色，一边欣赏珠江出海口的美景，一边感受莲花山之胜境，不失为一大享受。

Sometimes at the wharf of Panyu Lotus Hill, fishermen can catch some sea crabs, such as big red crabs, big in size and delicious in taste. One can select and enjoy nice crabs with several close friends at the wharf or Seagull Island, enjoying the beautiful scenery of the Pearl River estuary and the Lotus Hill.

荔枝，不只是佳果
Lychee, More Than a Good Fruit

公元745年，炎夏，南雄官道尘土飞扬，数匹唐朝一等一的快马奋蹄疾奔，穿越南岭。马背上，是南粤人种植的荔枝佳果；马蹄下，则是南粤人生凿开通的天堑通途，目的地是2100公里外的首都长安（现陕西西安），足足七天的日夜兼程，换来的是杨贵妃的微微一笑。

In the summer of the year 745, a number of best horses were rushing through an ancient road of Nanxiong to the destination, Chang'an, the capital which was 2100 kilometers away. On the horses were lychees planted by Nanyue people; under the horses were the thoroughfares built by them. The arduous journey of seven days and nights was to win a smile of Imperial Concubine Yang.

杨贵妃的特权在历史长河中是留不住的，却成就了"一骑红尘妃子笑，无人知是荔枝来"的千古绝句，以及一种叫作"妃子笑"的荔枝品种。

The privilege of Imperial Concubine Yang could not be retained in the long history, but it leaves the eternal verse of "the concubine smiles at a dashing horse, and no one knows it is lychees coming" and a strain of lychees called "Concubine Smile".

你可以想象,杨贵妃丰腴白亮的兰花指捏起晶莹圆润的果肉,相互映衬成了一种专属于古代皇家的美学。
You can imagine that Imperial Concubine Yang picked up the crystal and mellow pulp of lychee, which is a exclusive scene of ancient royal families.

广州人吃荔枝可不是这样的。
However, natives of Guangzhou don't eat lychees like this.

身在荔枝湾,感受着"出郭先经晚景园,半塘南岸果皆繁。三山大石红相望,熟到陈村并李村"的盛景。即便是在夏日艳阳高照之下,旧时的人们亦无惧色,因为荔枝树排列整齐,远观似红云一片,枝繁叶茂的荔枝为它的追捧者提供天然的庇荫。现今的荔枝湾经过排污整治、升级改造,其景更胜往日,此为后话,暂且不提。
At Lychee Bay, one can feel the magnificent scenery described in the verses: "Walk through the Wanjing Garden, seeing that the south bank of Bantang is full of fruits. The three mountains are all red from village to village." Even in the blazing summer sun, people weren't worried about the sunlight.Because numerous lychee trees were closely planted and looked like red clouds from afar, forming

natural shades for lychee lovers. Now the Lychee Bay, after sewage treatment and renovation, is even better than the past.

人们将荔枝红色的果壳一圈圈剥开，晶莹的果肉一点点露出真容，仿佛在掀开美丽新娘的红盖头，待白嫩的果肉露出大半，便可轻挤到嘴里。荔枝初入口时顿感一阵清凉，牙齿切入时甜美的汁液流入喉舌、沁入心扉。如此动作和感受循环往复，人们在闲话家常的同时，三五斤荔枝轻而易举就被"消灭"了，"日啖荔枝三百颗"的名句大抵就是这样来的吧。虽不一定有杨贵妃的美感，但却平添了广州人身上少见的一股豪气。当然，这"三百颗"只是文学修辞上的虚数，因为大文豪苏东坡再怎么样也不可能一天之内吃完这 10 多斤佳果。

People tear the lychee peel in circles, and the crystal pulp displays itself bit by bit, as if uncovering the red veil of a bride. When much of the pulp is exposed, it can be squeezed into the mouth. Once in the mouth, people feels cool, And when bitten, its sweet juice flows into the tongue, the throat and the heart. Several kilos of lychees are easily consumed when people repeat such actions and feeling while chatting at home. The famous words of "eating three hundred lychees a day" are probably like this. Such a way of eating lychees might not be as elegant as Concubine Yang, but enhances the sense of heroism not commonly seen among natives of Guangzhou. Of course, "three hundred" is just an imaginary number in literary rhetoric, otherwise how could Su Dongpo, the famous literary giant eat over ten pounds of lychees a day?

荔枝一般在5—7月成熟,而最早"报到"的荔枝品种是"三月红"(这里说的"三月"是农历,其实就是公历四五月)。但广州人认为吃荔枝的最好时节是夏至,一般是6月21日或22日这两天中的一天,因为夏至正是荔枝收获的旺季。

Usually lychees mature between May and July, the earliest one being "March Red" (here March is in the lunar calendar, which is actually April or May in the solar calendar). But natives of Guangzhou think the best time to eat lychees is Summer Solstice, because it is the harvest season of lychees.

岭南地区盛产荔枝,有统计数据显示,广东的荔枝种植面积有29.4万平方米,占全国的二分之一。而开创岭南种植荔枝之先的正是广州的增城。广州人在吃的方面从来不缺乏想象力,荔枝数量既多,食客们就开始动脑筋了。

The Lingnan Area is rich in lychees. Statistics show that the lychee planting area in Guangdong is about 294 thousand square meters, accounting for a half of the whole country. The first place to plant lychees is Zengcheng of Guangzhou. As natives of Guangzhou always have rich imagination in eating, when there are so many lychees, the foodies start exploring ways of eating them.

要不先用荔枝做点甜品吧,于是就有了果冻一般的贵妃糕。将新鲜采摘的荔枝去核,枸杞加水泡开,鱼胶粉与白糖放在一起熬制搅拌,然后将荔枝、枸杞和熬制好的鱼胶粉浆一并倒入模具,

天时 | 第二节 | 食得新鲜 ◆ 夏
Right Time | Section II | Eating Freshly ◆ Summer

■ 荔枝红茶凉糕

放入冰箱里冷冻 40~50 分钟。清甜可口的荔枝,搭配上冰冰凉凉的果冻,吃下去自然让人透心凉。

  Why not make some lychee desserts? there is jelly-like "Concubine Cake". Pit the newly picked lychees, soak medlars with boiling water, boil and stir fish gelatin powder and sugar, then put the mixture into molds and freeze it for 40 to 50 minutes. The sweet lychees paired with icy jelly taste so cool.

  用新会红柑做盏,盏中盛清汤,甘蔗削成条,插上荔枝肉和牛肉球,似孩童喜欢的棒棒糖。吃法有讲究,先用小吸管喝红柑中的清汤,再享用甘蔗荔枝牛肉球。

Red oranges from Xinhui are used as containers, in which there is clear soup. Sugarcanes are cut into small sticks, on which lychees and beef balls are strung together, like a lollipop. The eating method is exquisite. One has to drink the soup with a straw before enjoying the sugarcanes, lychees and beef balls.

还有粤菜师傅往荔枝里酿鲜虾。粤菜师傅将打好的虾胶酿入去核的荔枝果肉中，取虾胶和百合片一揉一拧，便成了莲花一般秀美的底座，正好将酿荔枝稳稳固定，俨然百花绽开。大口啖下，荔枝肉和虾肉在汁液的刺激之下分外清爽！

Some Cantonese cooks add fresh shrimps to lychees. The cook stuffs the pitted lychees with shrimp paste. The paste and lily cloves are made into lotus-like bases for lychees, so that they look just like blooming flowers. The lychee and shrimp taste so good and juicy.

当然还少不了广东人最爱的鸡，用荔枝、椰子和鸡肉共煮一煲，荔枝的清甜自不必说，椰子与鸡本就能烹出清甜的汤，三者"确认过眼神，就是对的味"。这鸡煲不加一点水，用4个椰青的汁液做汤底，加上雪耳、老椰肉、珍珠马蹄、文昌鸡和400克荔枝肉煮上两分半钟，熄火后再焖30秒。正确的食用方式是先喝上几口清汤，再开吃。

Of course, Cantonese favorite chicken is necessary. Lychees, coconuts and chicken are made into a soup. It is just the right taste with sweetness and freshness. No water is added to the soup. Juice of four coconuts serves as base, then snow fungus, coconut, water chestnuts, Wenchang chicken and 400 grams of lychees are added. It is cooked for two and a half minutes and simmered for 30 seconds. The right way of eating it is to sip several mouthfuls of soup before eating the ingredients.

荔枝吃法之丰富如是,已远远超出其水果的"人设",即便是你身在异国,也无须为品尝不到荔枝而惆怅,荔枝蜜、荔枝果酱已经流传到海外,独特的"岭南甜"早已绽放在国外的餐桌上、甜点里。

■ **荔枝饮品:贵妃醉清饮**

■ 荔枝菌————（高敏华摄影）

There are so many different ways of eating lychees, making them much more than a mere fruit. Even people living abroad don't need to feel sad for not being able to eat lychees, because lychee honey and lychee jam are already sold overseas. The unique Lingnan fruit has made its way to the table and the desserts in foreign countries.

你以为这样就完了吗？不、不、不。夏末午夜的岭南乡村里，农民打着小灯，在荔枝林中寻觅一种珍贵的食材——荔枝菌。荔枝菌其实不过是鸡枞菌的一种，但其珍贵之处就在于，荔枝菌只能在特定的气候、环境中生长，且出现时间极短，如同昙花一般金贵。至于为什么荔枝菌一般只在荔枝树下觅得，并无权威答案，农民们普遍认为，荔枝菌的生长有赖于荔枝树下的白蚁穴，白蚁的分泌物促进了荔枝菌的成长。

Do you think this is all about it? No, no, no. In Lingnan villages at summer nights, peasants, with little lamps in their hands, are looking for a precious food material, lychee fungus. Though it is only a kind of collybia albuminosa, it is precious because it can only grow in certain climate and environment for a very short period of time, just like night-blooming cereus. Though there is no authoritative answer to the question why the lychee fungus can only be found under lychee trees, the peasants believe that it all relies on the termites which live under the lychee trees. The enzyme secreted by the termites boosts the growth of the lychee fungus.

荔枝菌色泽洁白，菌身较硬，吃起来只觉细嫩爽口，菌味清淡，鲜美中带着丝丝泥香，让人试过一次便对其牵肠挂肚。荔枝菌的保质期非常短，在0℃~4℃的冰柜里也最多只能保质3天，实际上，一般两天后，荔枝菌就"开伞"，再过12个小时，其营养精华则全部流失，这样的荔枝菌则沦为次货，广州的食客们对荔枝菌从趋之若鹜到嗤之以鼻，只需要3天，变脸速度这么快，正好说明了广州人对"不时不吃"的信仰有多深。

The lychee fungus is white and hard. It tastes tender and refreshing, with a hint of mud. Whoever has tried it once will remember it forever. The lychee fungus has a very short shelf life, at most three days

■ 荔枝菌（高敏华摄影）

with temperatures between 0℃ ~4℃ in the refrigerator. In fact, usually after two days, it begins to spread spores. After another twelve hours, its nutrition has scattered. Such lychee fungus is regarded as inferior and scorned at by Cantonese foodies in Guangzhou. People's attitudes toward it change completely within just three days, revealing how much they believe in "eating at the right time".

　　虽然白蚁是公认的害虫，但也成了广东厨师的好"帮手"，这样物尽其用的方式正体现广州人与自然和谐相处的生活理念。广州人甚至还不放过荔枝木——荔枝树的树枝。难道还吃树枝吗？当然不是，荔枝木是用来做燃料的，熬粥，做烧鸡、烤肉，用荔枝木就是"零舍唔同"（广府俗语，意为"特别不同"）。全因荔枝木质地特别结实，干燥耐烧，燃烧时烟不大，持久而温柔的火力，甚至可以将荔枝的香气从树枝中逼出。

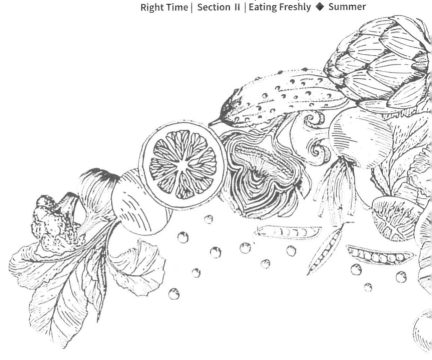

Though termites are inarguably insect pests they have become a "good aide" of Guangzhou chefs. Making full use of things in such ways speaks of the pursuit of harmonious coexistence of man and nature by natives of Guangzhou .They even don't spare the branches of lychee trees. Can the branches be eaten? Of course not. They are used as fuels for cooking porridge, roasting chickens and meat. Using lychee branches surely makes a difference. Because the lychee wood is particularly strong, dry and long lasting with few smokes when burning, the constant gentle fire even forces the fragrance of lychee out of the wood.

这香气，就像红线女唱《卖荔枝》的那股高腔滑音，绵长袅袅，在一代代岭南人的味觉中传递。

This fragrance is just like the high-pitched portamento sung by Hung Sin-nui in the Cantonese Opera *Selling Lychees*, long and lingering, passing from generation to generation among Lingnan people.

第三节 | 食得新鲜 ◆ 秋
Section III | Eating Freshly ◆ Autumn

秋风乍起时，
只追禾虫肥

When the Autumn Winds Blow,
Rice Worms Are At Their Best

禾虫出广州，禾熟时有之，长数尺，土人断之，亦复成形，形如马蝗，以治馔甚鲜。

——《岭南风物记》

Rice worms(Tylorrhynchus)exist in Guangzhou and appear when rice matures. It is several feet long. When local people break it, it restores itself again. It is shaped like a leech. It is very delicious to eat after being cooked.

——*Records of Lingnan Scenery*

农历四月十七前后，禾熟季节，水质优越的稻田里，农户们正等待着潮退后的"邂逅"。让农户翘首以盼的不是邻村的姑娘，而是稻田里胖乎乎的早造禾虫。这是珠三角一带备受好评的低脂高蛋白传统可食虫。潮水一退，它们从田中流出，涌入河中，农户即刻用簸箕接住，满筐收获，大喜。

Around the 17th day of the fourth lunar month is the time when the rice matures. In paddy fields where the water quality is superior, peasants are waiting for the "encounter" after the tides ebb. It is not the girl from nearby villages that they are waiting for, but the chubby rice worms in the paddy fields. They are low-fat, high-protein traditional edible worms in the Pearl River Delta. As soon as the tides ebb, they flow out from the fields and into the river. The peasants catch them with dustpans and rejoice at their gains.

禾虫身娇肉贵，生长环境容不得半点污染，一点农药也不能忍。因此，它被赋予一个神奇的身份：环境转好的"见证"。环境差，禾虫产量就少；环境好，禾虫产量就增加。广州番禺、南沙等地的稻田都是禾虫栖身之所。

Rice worms are so delicate that they cannot suffer any pollution or pesticide in their living environment, so they are endowed with a special identity, the "testimony" of an improved environment. When the environment is bad, the production of rice worms is small, and vice versa. Panyu and Nansha of Guangzhou are home to rice worms.

禾虫一年两造。农历四月十七前后的禾虫为早造,因农历四月十七是民间金花诞,故早造的禾虫有"金花虫"之称。此时的禾虫是上一年的老虫冬眠后出来的,皮稍微厚实些,口感相对粗糙,但无碍鲜美度。农历八月出没的晚造禾虫则更为上品,它拥有一个名号"大造虫",可想而知禾虫是多么肥硕,丰腴多浆。

Rice worms mature twice in a year. The early ones mature around the 17th day of the fourth lunar month, which is the Golden Flower Anniversary. Thus they are also called "golden flower worms". These worms come out from hibernation in the former year. Their skins are thicker and they taste a little rough, but they are by all means as fresh as ever. Those appear in the eighth month of the lunar calendar are called "late worms" or "big worms". Just imagine how strong and juicy they are.

明末清初著名学者屈大均在《广东新语》中记载:"夏秋间,早晚稻将熟,禾虫自稻根出,潮长浸田,因乘潮入海,日浮夜沉,浮则水面皆紫。"坊间有一句话叫"禾虫过造恨唔翻",更加印证了广府人"不时不食"的态度。

In the late Ming and early Qing dynasties, the famous scholar Qu Dajun recorded in *New Canton*, "Between summer and autumn when the early rice and late rice are about to mature, the rice worms come out from the roots of rice. When the tide immerses the field, the worms flow into the sea. They flow on the surface in the day and submerge at night. When they are floating, the water turns purple." There is a saying among the locals, "It is a regret to miss the time when the rice worms mature." This reflects Cantonese attitude of "eating at the right time".

别看禾虫只有小小的身躯,却大有益处。《本草纲目·拾遗》中称其"云食之补脾健胃"。这小家伙全身无骨,只一层薄衣裹浆液。美食家冼悦华是番禺大石人士,他最怀念孩提时吃过的禾虫饼。据他描述:那是在八月时,天尚热,人们将一片芭蕉叶放在禾虫出没之处,禾虫纷纷爬到蕉叶上唞凉(广府俗语,意为"乘凉")。因蕉叶的温度低于周边,不一会儿工夫,蕉叶上集合了一大批禾虫。然后将蕉叶放在太阳下晒,禾虫一受热就爆浆,浆液糊成一片,久晒成饼状,是为"禾虫饼"。再拿去一蒸,哇!美味到几碗白饭都能吃下。在中山当地更能吃上禾虫宴,包括禾虫焖柚皮、禾虫栗子焖烧腩、腌咸禾虫蒸腊肉等。

Though the rice worms are small, they are of great value. In *A Supplement to Compendium of Materia Medica*,

■ 蛋焗禾虫(石忠情摄影)

they are said to "benefit the spleen and stomach". The little thing has no bones but just a thin coat outside its flesh. Gourmet Xian Yuehua, a native of Dashi in Panyu, really cherishes the memory of eating rice worm cakes in childhood. As he describes, in August when the weather was still hot, a banana leaf was placed at the whereabouts of the worms. Because the temperature on the leaf was lower than its surroundings, the worms all climbed to it to enjoy the coolness. After a short while, a large number of worms gathered on the leaf. When it was put under the sun, the worms exploded with heat and turned into a paste, hence the "rice worm cake". The paste was then steamed. Wow! It was so delicious that one could swallow several bowlfuls of rice with it. In Zhongshan City, people can enjoy rice worm feasts, which include simmered rice worms with pomelo peels, stewed streaky meat with rice worms and chestnuts, and steamed preserved meat with salted rice worms.

石楼钵仔禾虫是番禺十大名菜之一，嫩滑可口，清香鲜美。新鲜禾虫在瓦钵中自饮生油，禾虫饮饱油后"自爆"，此时再加入陈皮、蒜片、红糖、胡椒粉、鸡蛋浆和炸油条，蒸熟后放在炉子上烤到焦黄，撒上花生碎与柠檬叶丝，美味得叫人毕生难忘。还可选择加蛋浆或不加。

Shilou Clay Bowl Rice Worms is one of the Top Ten Dishes of Panyu. It is tender, delicious and fresh. The live worms drink unboiled oil in the clay bowl. They "explode" after drinking. Then dried orange peels, sliced garlic, brown sugar, pepper powder, egg glaze and deep-fried dough are added. The mixture is steamed and then baked on the stove until turning brown. Chopped peanuts and sliced lemon leaves are added. The taste is so unforgettable. One can choose whether to add egg glaze or not.

　　你猜收拾禾虫难度指数最高的是哪一项技法？炒！身段柔软如它，稍微一用力必定全部断裂。怎么办好呢？充满智慧又酷爱禾虫美味的广州人想出了一个好法子——焯水。先用虾眼水焯过禾虫，再捞起晾干，下锅慢火炒熟。冼悦华还有另一个好法子：放冰箱急冻，将禾虫冻成"雪条"（广府俗语，意为"冰棍"），解冻后隔水蒸熟，下锅。加入时令凉瓜、红糖和豆豉共炒，炒到禾虫干身、甘香即可，这叫作凉瓜禾虫。据说有人爱之若狂，还发明了禾虫刺身的吃法，真可谓巧妙到了极致。

　　What do you think is the most difficult way of cooking the rice worms? Stir fry. The worms are so tender that they are broken with just a little force. What should be done? The intelligent and worm-loving natives of Guangzhou have come up with a good idea: boiling them first. The worms are boiled with water. After being scooped out, they are left to dry before being stir-fried with low fire. There is another good way suggested by Xian Yuehua: freezing them in the refrigerator into "ice sticks". After unfreezing, steam them before stir-frying them with some seasonal bitter melon, brown sugar and preserved black beans. When the worms become dry, the dish is ready to serve. It is called "rice worms with bitter melon". Some people love rice worms so much that they have even invented rice worm sashimi. What an artful awy of tasting this delicious worm!

第四节 | 食得新鲜 ◆ 冬
Section IV | Eating Freshly ◆ Winter

"虽迟但到"菜心味

Late but Right Choy Sum
(Chinese Flowering Cabbage)

有心栽花花不开，
无心插菜菜成片。
此味本只天上有，
广府却得时有品。
No flower blooms despite intentions,
Vegetables grow well accidentally.
They should belong to heaven,
But Cantonese taste them timely.

打油诗虽然是笔者生造的，但用来形容迟菜心的诞生却颇为贴切。"八仙过海，各显神通"。八仙中唯一的女性——何仙姑，就是广州增城人士，其擅长用各种仙草灵药，为人治病。然而，她大概没有预料到自己的一次大型义诊活动还会给广州人带来一种长久的食材——迟菜心。唐朝年间，天下大旱，很多人中暑而亡。何仙姑为解救世人苦难，将仙草结成草马，下界化身凉粉草，让人熬制成药，以解夏暑。草马在下凡时一不小心，身上多沾了一颗种子，还把它带到了凡间，和凉粉草生长在一起，变成了一颗清甜、多汁的野菜。传说那就是现在的增城迟菜心。

This doggerel is coined by me, but can properly describe the legend of the origin of Late Choy Sum. The only woman in the legend of the Eight Immortals, He Xiangu, was a native of Zengcheng. She was good at using various herbal medicines to cure people. However, never did she expect that one of her large-scale free clinics would bring to Guangzhou a long-lasting food—the Late Choy Sum. In the Tang Dynasty, because of drought, a lot of people died of heatstroke. In order to save the world from suffering, she made the immortal grass into a grass horse and after reaching the earth, turned it into jelly grass. So that it could be brewed to save the people from the heat. When the grass horse was descending, it accidentally brought a seed to the mortal world, which grew together with the jelly grass and turned into a sweet and juicy vegetable. This was said to be the origin of Zengcheng Late Choy Sum.

迟菜心一般是农历十月（即公历11月前后）播种，经过90～120天的生长期，冬至前后收获。虽传说何仙姑为迟菜心的播种人，但为迟菜心打造"优秀广告"的却是文人。明代一位秀才叫张贤，为了躲避战乱来到增城，当时正值寒冬，饥寒交迫之时，张贤看到路边种有一种高大的野菜，在冬日里显得青翠盈盈，随即采来烹食。张贤发现吃起来竟然液汁清甜、爽滑无渣，吃完后竟觉饥渴尽消、精神饱满。后来，张贤考上进士当了尚书，故地重游回到增城，得知当日充饥的野菜，竟是传说中何仙姑的草马带落凡间的种子生长出来的，顿时诗意大发，为该菜题了诗："菜心盈翠冬日迟，天赐丰饶胜奇葩。"

这句经典的"广告词",一举为增城菜心画龙点睛地添了一个"迟"字,更让增城迟菜心声名远播。

The seeds of this vegetable are planted in the tenth lunar month (around November of the solar calendar). It takes about 90 to 120 days to grow and matures around the winter solstice. It is Immortal He Xiangu that has brought the seed, but it is a scholar that has "advertised" it. In the Ming Dynasty, a scholar named Zhang Xian came to Zengcheng to escape from the war. It was the cold winter season. Zhang, cold and hungry, found a tall wild vegetable on the roadside. It looked so green in the winter sun. Zhang cooked it and found it so juicy, sweet and residue-free. After eating it, Zhang no longer felt hungry or thirsty. Instead he was full of energy. Later, after he became a high official in the imperial court, he visited Zengcheng again. He came to know that the wild vegetable that he had eaten was grown from the seed brought to the earth by the grass horse of Immortal He. Inspired, he wrote a poem for the vegetable, "So green is Choy Sum in late winter, heaven-blessed richness outshines exotic flowers". This classic "advertisement" highlights the lateness of Zengcheng Choy Sum, making it known far and wide.

秀才张贤与迟菜心的"冬日邂逅",成就了一款流传至今的靓食材。

Scholar Zhang Xian's "winter encounter" with the Late Choy Sum has turned it into a long-cherished food.

经历了冬季寒意洗礼的菜心长得特别高大，甚至能长至1米之高。而"体重"方面，一般菜心一颗不过一两多重，而茎肥叶厚的迟菜心一颗就可重达1斤，甚至可以重达3斤。

The Late Choy Sum, which has endured the harsh cold winter, grows to be as high as one meter. In terms of weight, an ordinary Choy Sum is just over 50 grams, but a Late Choy Sum can be 500 grams or even 1500 grams.

增城迟菜心一年才一造，比普通菜心收割都要迟，其清甜"虽迟但到"。冬天里一波又一波的冷空气并没有击垮它，反而让它将这些"伤害"化成糖分积蓄在身体里，越是迟到，积蓄的糖分越丰富；经历的寒风越多，越是能锻炼出脆皮嫩肉，霜冻之后才收割的迟菜心，完全可以称得上是"菜中极品"。

Zengcheng Late Choy Sum matures only once a year, much later than ordinary ones. It is late but right. The waves of cold air in winter do not destroy it. Instead, it turns these "harms" into sugar and accumulates it in the body. The later, the sweeter. The more cold winds it suffers, the better it tastes. Late Choy Sum harvested after the frost is considered "the best of all vegetables".

无论是个头还是上市时间，增城迟菜心都是南粤蔬菜品种中的"异类"，而这"异类"就好像是自然赐予广州人过冬的礼物。大家都知道，广州人一年四季离不开青菜，每到冬季青菜减产，广州人便惆怅不已。幸好有了迟菜心，填补了冬天的口福空档，让广州人在入冬后仍然能"争鲜"。当迟菜心的鲜甜绽放在舌尖之时，你定会如我一般感慨，这样的美味值得等待，大自然确实待广州人不薄。有趣的是，普通菜心空心是不好吃的，但增城迟菜心却是越空心越好吃。

As to its size or its time to the market, Zengcheng Late Choy Sum is an "alien" among vegetables in South China. However, this alien is just a gift endowed to natives of Guangzhou, who cannot live without green vegetables all the year round. When the production of vegetables declines in winter, they become so sentimental. Fortunately, they have the Late Choy Sum, which enables them to enjoy fresh vegetables even in winter. When savoring the sweet and fresh Late Choy Sum, one cannot help exclaiming that such delicacy is worth waiting for and that nature treats the people in Guangzhou so well. It's interesting that for ordinary Choy Sum a hollow stem means it is not good to eat, but for Late Choy Sum, the hollower, the tastier.

■ 迟菜心与腊肉是绝配

　　何仙姑的故事自然仅仅是神话了，但迟菜心在广州增城的历史却是真实的，特别是增城的小楼镇腊布村，早在20世纪中叶就开始有人种植迟菜心。然而，这"傻大个"的外形大概并不太符合广府人对"精致"美食的期待，一开始迟菜心并不太受欢迎，因此腊布村村民更倾向于种橙子、荔枝等水果。

　　The story of Immortal He is merely a myth, but the history of Late Choy Sum is more than real. In Labu Village, Xiaolou Town of Zengcheng, people started planting it as early as the mid-20th century. However, its big and clumsy appearance did not meet the expectations of the natives of Guangzhou for "exquisite" cuisine, so it was not popular from the very beginning. As a result, the villagers of Labu Village turned to plant fruit trees such as oranges and lychees.

　　但迟菜心似乎注定会在灾害后体现其价值。1999年冬天的一场特大霜冻，将村里许多果树摧毁，少有的几颗迟菜心却丝毫无损，其青绿更夺目，其清甜更甘润。迟菜心的价值终于被腊布人发现，并逐步开始大面积种植。

　　However, the Late Choy Sum was destined to show its value after disasters. In the winter of 1999, heavy frost destroyed many fruit trees, but the Late Choy

Sum remained intact and became even greener and sweeter than ever. Its value was found by the villagers who began large-scale planting.

既然大自然如此厚待广府人,广府人亦能以奇巧厨艺让迟菜心绽放新的光彩。有经验的厨师会从切菜开始就为迟菜心打造最合适的烹法,菜茎斜切,胸径横切,方能在炒制时凸显其清脆口感。

As nature prouides people in Guangzhou so many resources, they bring the charm of the vegetable to the full with brilliant cooking skills. Experienced cooks know how to choose the most suitable cooking method for it. For example, the stems are cut diagonally and the main parts transversely so that its crisp taste can be highlighted.

炒、煮、滚汤、白灼都只是等闲做法,新意者也不绝于"菜心界"。如将菜心做成春卷的馅料,晶莹剔透的粉皮里包裹着腊肠、菜心和黄瓜。明明是春卷,却得名"银装素裹",迟菜心诞生的冬天与广州人喜爱的春天,在此产生了某种微妙的勾连。而云吞这一广式小点,亦可成为迟菜心发挥的舞台。将鲜肉与迟菜心粒混合成馅,快速包成云吞,再用葱在云吞封口处打结。云吞煮熟后,再配搭上成颗迟菜心摆成树木造型,是为一道玉树藏福袋。

Stir-frying, boiling, soup-making or blanching are just ordinary cooking method. Innovations keep coming up. For example, it is made into the fillings of spring rolls. The crystal and transparent coverings wrap up sausages, Choy Sum and cucumber. Though it is apparently spring roll, it is named "silver snow", being named after the season winter when the Late Choy Sum grows and spring, the favorite season of natives of Guangzhou in a subtle way. Wonton, a Cantonese dim sum, has also offered a stage for Late Choy Sum. The diced vegetable and fresh minced meat serve as fillings wrapped into wonton, which is tied with shallots. After the wonton is cooked, it is paired with tree-like Choy Sum, so the dish is called "jade tree with blessing bags".

　　一颗迟菜心在寒冬遇上懂吃的广州人，以及巧手妙思的粤菜师傅，就如同在对的时间遇上了对的人，岂能不琴瑟和鸣、相得益彰？

　　A Late Choy Sum encountering the food-loving natives of Guangzhou and the clever and deft Cantonese cooks is just like meeting the right people at the right time. Aren't they mutually beneficial?

■ 迟菜心可以做成"卷"

炭火炉带来的
冬日暖意

Winter Warmth by the Charcoal Stove

绿蚁新醅酒,
红泥小火炉。
晚来天欲雪,
能饮一杯无?
Green bubbles on the newly brewed wine,
A Little red clay stove.
Snow at night,
Have a cup of wine?

  唐代诗人白居易在这首《问刘十九》的诗歌中,仅用寥寥数笔,便将寒夜中红泥炭火炉所带来的暖意渲染得淋漓尽致。很多广州人都曾有这样的经历:寒夜里,骑楼下,多则五六人,少则两人,围坐一桌,桌台中间摆一个炭炉,炭火熊熊燃烧,砂锅中的各种食材沸腾翻滚,香气便从盖子的缝隙中飘出,摇曳的香气与热量驱散了冬夜的寒冷。人多时,一边闲聊,一边等肉熟,时间过得飞快,仿佛眨眼就能开煲大吃;人少时,两个人对着炉子,对谈闲聊之间,煲里的肉也很快就熟了。

  In the poem "Inviting Liu Shijiu", the Tang Dynasty poet Bai Juyi vividly describes the warmth brought by a red clay stove with just a few words. Many natives in Guangzhou have such an experience: in a cold winter night in an arcade, as few as two or three and as many as five or six people sit around a table, on which there is a charcoal stove, with the flames burning and the food boiling in the pot. The smell of the food comes out from the seams of the lid. The swaying fragrance dispels the cold of the winter night. When there are many people chattering and waiting together, time flies quickly and in an instant the food is ready. Even when there are just two people whispering with each other, the meat is also ready quickly.

在广州，打火锅又称为"打边炉"。与别地的火锅不同，广州最传统的火锅使用的是炭炉和砂锅。如今，在老西关一带和海珠区的老街巷中，仍有不少店家保持着炭炉打煲的传统。他们选择无烟环保炭，一餐饭下来，身上也清爽无异味。冬日开煲，最常见的便是羊肉煲和牛腩煲。牛羊囤脂过冬，个个膘肥体壮，

肉质最是肥美,正好对上了广州人秋冬进补的习惯。一煲羊肉或牛肉是许多老广在冬日里最难以割舍的美食情结。特别是广式羊肉煲,更准确来说是羊腩煲,肥瘦相间的羊腩充满油脂和羊肉的甘香。

In Guangzhou, hot pot is also called "hot stove". Unlike that in other places, the most traditional Cantonese hot pots use charcoal and pottery pots. Now, some restaurants in the back streets of Xiguan and Haizhu still use the traditional charcoal for hot pots. Smokeless environment-friendly charcoal is used, so after the meal, no odor is left on the clothes. The most common hot pots in winter are mutton pots and beef pots. Cattle and sheep store fat in winter, so they all grow fat and strong and their meat is most delicious. It just coincides with the natives' habits of taking tonics in autumn and winter. A pot of mutton or beef is the most entrenched desire for many natives, especially a pot of streaky mutton. The mutton brisket tastes both fat and sweet.

虽说广州人口味偏清淡,但在吃羊肉时却也乐于选择"重口味"——酱爆羊。酱爆羊的酱是整道菜的精华所在,所用的材料大同小异,不外乎花生酱、海鲜酱、腐乳和柱侯酱,但每一位做酱爆羊的师傅都有自己独特的配方,通过不同比例的调配,酱料便呈现出各自鲜明的特色。加入各种食材的羊腩煲不停地翻滚,激发出羊肉特有的鲜香。即便不蘸料,以汤底的浓郁程度,羊腩煲都足够美味了。食罢羊肉,放入腐竹、冬笋等吸味一流的食材，延续了羊腩煲的鲜美。假如"肉过三巡"，你还"食有余力"，那必须来一包方便面，这是老广吃羊腩煲的标配。将面条放入煲中，等面条散开，颜色渐深，充分锁住羊肉鲜味，趁热吃一口，方才结束这场炭炉羊肉煲的暖胃之旅。

Though the natives of Guangzhou prefer lighter flavored food, when eating mutton, they also enjoy heavy taste—mutton dipped in sauce, which is the essence of the whole dish. The ingredients for the sauce are more or less the same, such as peanut sauce, seafood sauce, fermented soybean curd sauce and Chee Hou sauce. But every cook has his unique way of mixing the ingredients so that the sauces become distinct. The broth with various ingredients keeps boiling, stimulating the unique flavor of mutton. Even if no sauce is dipped, the mutton pot itself is delicious enough. After eating the mutton, add dried bean curd sticks and winter bamboo shoots which can best absorb the tastes, so they are the "best partners" of mutton. If you can still eat more, you must add a pack of instant noodles, a standard practice of mutton hot pots for the natives. When the noodles spread out and become darker in color, they have locked the freshness of the mutton. Take a mouthful of the noodles when they are still hot. Only then can the feast of warming up your stomach with mutton hot pots come to an end.

天时 | 第四节 | 食得新鲜 ◆ 冬
Right Time | Section III | Eating Freshly ◆ Winter

■ 牛杂煲（王维宣摄影）

对于"无鸡不成宴"的广州人来说，在炭炉打煲这件事上，鸡也是必须有的。煲靓鸡煲的关键是鸡要新鲜，现点现做，更多的时候是以清汤打底，不加过多的酱料，以求最大程度保持鸡肉原有的鲜味。现在，鸡煲的"内涵"越来越多元化，除了常见的暖胃驱寒的猪肚鸡煲外，海鲜也和鸡肉成了好搭档，例如鲍鱼鸡煲、九节虾鸡煲、花胶鸡煲等，味道更加鲜甜。和三五亲友天南地北地聊天，吃着嫩滑的鸡肉，再搭配两瓶珠江生啤，满足地打个饱嗝，这大概是广州冬天最"叹"（广府俗语，意为"享受"）的体验了。

To natives of Guangzhou whose feasts cannot do without chicken, chicken is a must for hot pots as well. The key to a good chicken hot pot is that the chicken must be fresh and cooked on the spot. Only clear soup is used without too many ingredients to maximally keep the freshness of the chicken. Now, the ingredients for chicken hot pots become more

■ 啫啫鸡煲（王维宣摄影）

diversified. Pig stomach and chicken hot pot that warms up the stomach is popular in restaurants. Seafood has also become good partners for chicken, such as abalone and chicken hot pot, nine-section prawns and chicken hot pot, fish maw and chicken hotpot, all of which are even fresher and sweeter. It is probably the most enjoyable experience in winter to chatter with several friends and relatives, eating tender chicken, drinking two bottles of Zhujiang Draft Beer, and belching with satisfaction.

第五节 | 食得健康
Section V | Eating Healthily

饭菜如舟,靓汤如水
Rice is Like a Boat While Good Soup is Like Water

岭南之地,在古时通常是被罢黜官员流放之地。那一长串流放官员的名单,用"名人辈出"来形容也丝毫不过分。

In ancient times, Lingnan was usually a place for dismissed and exiled officials. It is not at all exaggerating to describe the long list as an award-winning list, on which most are celebrities in Chinese history.

刘君锡也在这份名单上,虽然算不上什么大腕,但若论健康长寿,他绝对是数一数二的。刘君锡被贬至岭南(古时所称"岭南"包括今天的广东、广西和海南等地),深受瘴气、湿气之苦。某次在广西桂林的游历,改变了刘君锡的命运——他遇到了百岁道人刘仲元,两人一见如故。最关键的是,刘仲元交给刘君锡一个养生方子,名曰"养气汤方",并嘱咐他说,每天清晨早起,梳洗完毕后,空腹先服用此汤,可保你一天无事,天天如此,便可终身无病。刘君锡遵照道人嘱咐,服用此汤后,虽在岭南数年，竟免除了瘴气之患。刘君锡后来回到了家乡，常年服用此汤，很健康地活到了90岁。

Liu Junxi, ranked among the top for his health and longevity, was not well-known at his time. When in office, he was degraded and banished to Lingnan (including Guangdong, Guangxi and Hainan in ancient times), suffering from diseases caused by miasma and damp. But after encountering a hundred-year-old Taoist priest called Liu Zhongyuan on his tour to Guilin in Guangxi, his life was changed. They became intimate at the first meeting. More importantly, Zhongyuan offered him a soup prescription for invigorating qi(a Chinese medical term that refers to the functional force of metabolism), and instructed him to take it on an empty stomach after getting up and cleaning up in the morning. Taking the soup every day, Junxi would be protected from illness for life. Surprisingly, thanks to the soup taken every day in those several years, Junxi kept a healthy life from illness caused by miasma in Lingnan. Later when he went back to his hometown, he still took the soup every day, so he lived healthily to his nineties.

这大概就是最早的岭南养生汤方了。如果你想知道这"养气汤方"里到底藏着什么"神药"，也可以找"养气汤方"的石刻拓本一解好奇，它此刻正静静地躺在广州中医药大学的中医药文化博物馆里呢。

■ 一品养生汤

It may be the earliest soup prescription for preserving health in Lingnan. If you want to know more about the prescription, you can read the ink copy of its stone inscription to satisfy your curiosity, which is now stored in the Museum of Traditional Chinese Medicine Culture in Guangzhou University of Chinese Medicine.

石刻拓本终会模糊,但喝汤的习惯却随着历史的演进愈加清晰,并与现代营养学产生了某种联系。无论这石刻"养气汤方"记载的是什么,也无论这汤方是否真的如此神奇,岭南人通过养生靓汤适应客观环境、改善健康的确有渊源。

The ink copy of stone inscriptions may fade, but the habit of taking soup has become more distinct with

■ 大枣养颜汤（钟达文摄影）

the evolution of history, and to some extent, has developed some connection with modern nutriology. No matter what the prescription records or whether it is that magical, it cannot be denied that people living in Lingnan do have the tradition of taking the soup to adapt to the natural environment and improve their health.

　　岭南的天气，归结起来就是两个字：湿热。一个普通广州人未必能跟你解释得清楚什么叫作"湿热"，但说到"祛湿""清热"的食材，却几乎能如数家珍，例如：陈皮、砂仁、五指毛桃、木棉、淮山、薏仁……全因为这个"湿热"，刚到岭南的人有时会发现自己"周身唔聚财"（广府俗语，意为"浑身不舒服"），但去医院检查却发现什么病都没有。既然没病，自然也就不能无缘无故吃药。于是，聪明的广东人顺手就在身边找到了一些药食同源的食材，这里说的"顺手"还真不是笔者"顺手"写下的——据统计，岭南药用资源有 4500 种以上，占全国药用资源种类的36%，其中陆地资源中植物类约有4000种，如此丰富的药用资源，简直就是"老天爷赐汤喝"。广府人只需要把熬制中药汤剂的过程略为改良，就能煲出更适宜家中烹制和食用、更容易让"似病非病"的人"痊愈"的养生靓汤。

  Damp and hot can be used to describe the weather in Lingnan. An ordinary native of Guangzhou may be unable to explain clearly what "damp and hot" really means, but he or she is very familiar with the cooking ingredients that are good for clearing "damp and heat in the body" that are caused by the special Lingnan weather, like Chenpi (sun-dried tangerine peel), fructus amomi, radix fici simplicissimae, ceiba, Chinese yam and coix seed. People new to Lingnan would feel uncomfortable because of the weather. But they would be told that they are so healthy that they don't need to take any medicine when seeing a doctor. Therefore, intelligent Cantonese have begun to seek some ingredients near at hand that can be used both as medicine and food, which is not made up by me but exactly put on records. Statistically, the species of medical resources in Lingnan are to more than 4500, taking up 36% of the whole country. Among the over 4500 species, around 4000 species are plant resources on land. Such plentiful resources should be regarded as a blessing from nature. So long as Cantonese improve the procedure of decocting, healthy soup will be more conveniently made at home and

■ 松茸瘦肉汤

help "cure" those suffering uncertain discomfort.

　　说到广府汤水，人们自然会想起老火靓汤。所谓老火靓汤，指的是用文火长时间熬煮的汤。然而，广府汤水并不等于老火靓汤。

　　Speaking of Cantonese soup, long-stewed soup does cross people's mind. This kind of soup is made by low and slow heat for a long time. However, it cannot completely represent Cantonese soup.

　　广府汤水的烹制方式有"煲""炖""滚"等类型。其中，老火靓汤的烹饪方法一般为煲；炖汤是将汤料放置炖盅内隔水炖，冬季多用；滚汤是以锅或铁镬为器皿，武火滚至刚熟即可，为夏季多用。可见，所谓老火靓汤不过是广府汤水中的一小部分，更何况，随着电子炖盅和高压锅等"高效"炊具的普及，即便是"煲汤"，也未必就真会熬上三四个小时，熬制一两个小时更为常见。

Cantonese soup can be cooked in different ways, like stewing, simmering and boiling. Stewing is usually used for long-stewed soup. Simmering is to place the soup in a container and simmer it in water, which is frequently used in winter. Boiling soup is mostly cooked in summer by high heat using a pot or iron braiser. It is thus clear that the long-stewed soup is just a small part of Cantonese soup. Moreover, with the popularization of efficient electric steaming pots and pressure cookers, one or two hours will be more common than three or four hours for a long-stewed soup.

不同时节选择不同食材，符合中医顺应时令养生的原则，例如：春天湿气重，所以煲鸡骨草炖猪横脷、五指毛桃薏米煲瘦肉；夏天湿热为主，就煲消暑清热解毒汤；秋天一般煲润肺润燥汤，如南北杏菜干猪肺汤、无花果炖猪骨；冬天大多煲进补汤水，如冬虫夏草瘦肉汤。广府人根据自身体质和配合外在环境对人体进行调理，以求达到天人合一、阴阳调和的状态。广府汤水食疗相信"人如其食"，同时也多与现代营养学暗合。根据时令来煲汤，并不是玄学，分明是穿越千年而来的中西医学文化精华嘛。

Different ingredients should be chosen in different seasons, a principle advocated by traditional Chinese medicine, which encourages people to cultivate health in conformity with the seasons. For example, damp weather is commonly seen in Lingnan's spring. In this case, pig spleens stewed with abrus cantoniensis hance, and lean pork stewed with radix fici simplicissimae and coix seed will be preferred. In the damp and hot summer, soup

■ 冬虫夏草瘦肉汤

for relieving summer heat and internal fever would be better. In dry autumn, soup for nourishing the lung like pig lung soup with apricot kernel and dried vegetables, and pig bone soup with ficus carica would be chosen. In winter, Cantonese usually cook pork soup with Chinese caterpillar fungus for nourishment. Nursing personal health according to the physical environment to reach the balance of yin and yang does conform to that principle. Cantonese dietary therapy is implicitly in line with modern nutriology. Cooking health soup according to the seasons is not metaphysics, but rather, an essential part of Chinese and Western medical culture which has lasted for thousands of years.

广府汤水不仅保健，还能"办成事"。唐宋年间的广州城就已有贸易传统，20万中原人、南越人与外商聚居于此，十分热闹。据记录唐代岭南风情的史籍《岭表录异》记载，南越人家特别流行喝"不乃羹"，谁要在生意场上请人吃个饭，宴席上必须有这一道不乃羹。不乃羹是用羊肉、鸡肉、鹿肉与骨头一起煲成，肥浓鲜美，上菜的时候，肉捞出来弃之不用，只以汤待客。一大盆汤摆在席上，大家用一个大勺子轮流取饮。之所以命名为不乃羹，意思是说，喝了这道汤，就没什么事办不成。你看，古人对汤的嗜好，与现在如出一辙，没准我们现在"只喝汤不吃肉"的习惯，也是从那时候传下来的呢。

Cantonese soup promotes not only heath but also business as well. As early as the Tang and Song dynasties, Guangzhou had the tradition of trade. Two hundred thousand traders from the central regions, Nanyue and foreign countries settled down and developed business in Guangzhou, creating a prosperous scene at that time. It's recorded in an ancient book of Lingnan customs in the Tang Dynasty called A Miscellany of Lingnan Customs that a vital soup named "Bu Nai Geng" was popular in commercial feasts in Nanyue culture. This course was made by stewing mutton, chicken, venison and bones together, which was delicious with a rich taste. When serving, the meat would be discarded, leaving the meatless soup in a big bowl for the guests, who would use the same big spoon to taste in turn. The name "Bu Nai Geng" means that people who have tasted the soup will succeed in making deals. The preference of soup in ancient times is exactly the same as now. It can probably be speculated that the tradition of drinking the soup instead of eating the meat has developed since that time.

这不乃羹不仅是"商场润滑剂",甚至还是"官民沟通"的一招。大宋官方有一项"德政",凡是节庆日,官府园林都要向老百姓免费开放。官府的仆人在其中还找到了商机,他们会在园子里摆个饮食摊,逛园子的百姓只要花一点钱,品尝一番不乃羹,也算是过把达官贵人的瘾吧。

"Bu Nai Geng" helped promote not only businesses relations, but also communication between officials and citizens. In the Song Dynasty, a benevolent rule released by the government gave citizens free access to gardens in official institutions in holidays. Servants working there saw it as an opportunity to earn money. They would sell "Bu Nai Geng" to visitors, satisfying their desire of tasting the soup as VIPs.

连官民沟通也都在一碗汤里,也无怪乎清代美食家李渔在《闲情偶寄·饮馔部》中如是说:"饭犹舟也,羹犹水也;舟之在滩,非水不下,与饭之在喉,非汤不下,其势一也。"在中国的文化传统中,舟和水的比喻通常用在君臣之间,但广府人却把这"国家大事"化作一碗汤、一席饭,岂不妙哉?人皆以为这是广府人的务实，甚至斥之为"短视"。然而，你又可否想过，"君臣之事"早已经被扫进历史的故纸堆中，而"养生靓汤"却千年来越喝越有滋味。哪个更悠长，哪个更"短视"，岂非一目了然的事么？

Even a bowl of soup can help communication between officials and citizens. It is not surprising to read that "rice is like a boat, while soup is like water. A boat can't move without the push of water, just as rice can't be easily swallowed without drinking soup", which was written by Li Yu, a Cantonese foodie-born in the Qing Dynasty, in his Sketches of Idle Pleasure: Food and Drink. In traditional Chinese culture, the comparison between a boat and water is usually used to demonstrate the relationship between an emperor and his officials or citizens, which is of national significance. Isn't it amazing to solve the state affairs through a meal or even a bowl of soup? People regard this way as a practical style of Cantonese, or criticize it as shortsightedness. However, feudal society has long passed. In modern society, soup for healthy life will be more and more popular. Isn't it clear which is long-lasting and which is shortsighted?

凉茶的甘苦之道
The Taste of Herbal Tea

讲凉茶,就要先讲两个年轻人的故事。
Speaking of herbal tea, the stories of two young men should be told first.

晋朝时,医药家葛洪游历到了岭南一带,正碰上瘴疠横行。葛洪不躲不避,干脆潜心研究岭南各种温病医药,用中草药煎水制出一批茶汤,它被认为是最早的凉茶方子。
During the Jin Dynasty, malaria (communicable subtropical diseases) broke out when pharmacist Ge Hong travelled to Lingnan. In this case, he didn't elude but stayed in Lingnan instead, studying various kinds of herbs. Finally he found out the earliest prescriptions of herbal tea, a kind of soup of herbs boiled with water.

是年,葛洪 22 岁。
At that time, Ge Hong was 22 years old.

时光荏苒、光阴似箭。千年过去了,到了清代道光年间,瘟疫肆虐岭南,另一个年轻人——王泽邦将岭南生长的凉性植物按照一定原理配成一剂良方,精心煎煮。而王泽邦的配置原理正是来源于葛洪的配方。1828年,王泽邦在广州城十三行的靖远街开设了第一间凉茶铺,向往来商客出售水碗凉茶。后来,这家店得名"王老吉"。

Time flies. During the reign of Emperor Daoguang in the Qing Dynasty, a pestilence was raging in Lingnan. Another person named Wang Zebang made up a prescription of cold plants in Lingnan, which conformed to the same theory as Ge Hong's, and boiled it painstakingly. In 1828, he set up a shop in Jingyuan Street in Guangzhou Thirteen Hongs, selling herbal tea in bowls to visitors and businessmen. Later, the shop was named *Wong Lo Kat*.

是年,王泽邦15岁。
At that time, Wang was 15 years old.

葛洪和王泽邦,一位是"青年医学家",一位是"青年创业者",两个人一段跨越千年的合作,开辟了凉茶在岭南地区的历史。而在他们之后,凉茶的历史注定要超过千年。

The cooperation between a young pharmacist and a young entrepreneur across a thousand years opened up the history of herbal tea in Lingnan. Since then, the history of herbal tea is destined to exist more than a thousand years.

两个半人高的葫芦状大铜壶,一列青花粗瓷大碗,墙上挂着火炭母、水瓜壳等干草药物,还有几张待客的木凳,这就是清末一般凉茶铺的全部家当。看似简单,但一家凉茶铺的装潢细节仍可体现岭南人的体贴入微。一字排开的各式茶煲、茶碗,对症,更重要的是对时令;碗口用圆玻璃片盖上以保清洁和温度,直白地告诉不懂行的顾客：凉茶的"凉"，不是茶的温度，而是解热毒的"凉"。至于"惯入深山寻百草，隐于闹市卖凉茶"的常用对联，更像是一个充满禅意的广告。

An ordinary herbal tea shop in the late Qing Dynasty was roughly decorated with two big waist-high gourd-like copper pots, a set of big bowls with blue and white porcelain, with dried herbs like Chinese knotweed and melon sponges hanging on the wall, as well as several wooden benches for customers. The seemingly simple decoration reflects the great consideration of Lingnan people through the details in the shop. Various kinds of pots and bowls lining up were used for different symptoms, and more importantly, for different seasons. The bowls were covered with glass to keep the herbal tea warm and clean, directly showing to those customers knowing little about herbal tea that the tea (literally translated as "cool tea"), instead of serving as a cool drink, is to cool and relieve the internal heat. The couplet, "seeking herbs in remote mountains, selling herb-tea in busy streets", commonly seen in herbal tea shops, is more like an advertisement with Buddhist significance.

　　而如今，在粤港澳地区，"凉茶铺多过米铺"、凉茶铺"梗有一间喺左近"（广府俗语，意为"总有一间在附近"）的现象往往让外乡人难以理解。

But nowadays, it is difficult for non-natives to understand why the number of herbal tea shops exceeds that of rice shops and why they can exist everywhere in Guangzhou, Hong Kong and Macao.

　　凉茶为何能在百余年间深深植入岭南人的生活中呢？

Why can herbal tea become a necessary part of Lingnan people's daily life for more than one hundred years?

　　重要的是凉茶在1840年前后发生的变革。其时，凉茶被第一次制作成茶包，从此凉茶不再一定需要一个水碗，广州人可带着茶包，便可以"一茶走天涯"。自此，凉茶便有机会随着侨民"冲出广州，走向东南亚"。

The great changes taking place around 1840 did play an important role. At that time, herbal tea shops no longer just sold herbal tea in a big bowl, but rather, began to sell tea bags, which were convenient for Guangzhou citizens to take wherever they wanted. Since then, herbal tea has become popular outside Guangzhou thanks to Chinese nationals emigrating to Southeast Asia.

重要的是20世纪二三十年代前后出现的凉茶"走鬼"。一个个老头在肩膀上围些树叶,扮作"神农氏"模样,排起队伍,敲着锣鼓,齐声高唱:"神农茶,神农茶,发烧发热有揸拿(广府俗语,意为"有把握"),不妨买包神农茶。""走鬼神农氏"们走街串巷地"营销",也是当年的广州一景。

The vendors in the 1920s and 1930s also contributed a lot. They dressed up as Shennong (also known as the Emperor of the Five Grains) with leaves on their shoulders, lining up and beating the drums, singing together "Shennong tea, Shennong tea, no matter fever or heat, it can beat". The special marketing of those Shennong vendors from street to street became a significant scene in Guangzhou at that time.

然而,这些都不是最重要的。"平衡"才是凉茶成为广州延续千年的饮品的文化内因。

However, these are not the most important factors. "Balance" is the underlying cultural reason for herbal tea to become one of the most popular drinks in Guangzhou in the past millennium.

广州空气潮湿，容易滋生细菌，饮用祛湿清热解毒的凉茶，或防范感冒于未发之前，或抵御病毒于入侵之后，正是平衡人体与环境的妙方。

Since ancient times, the weather in Guangzhou has been damp and moist, which easily breeds bacteria. It is a wonderful way to balance our physical health and the natural environment by drinking herbal tea to clear internal damp and heat, before catching a cold or after the virus infection.

天时 | 第五节 | 食得健康
Right Time | Section V | Eating Healthily

广州人自古好吃，一天七顿只等闲，再加上"一个荔枝三把火"，如此一来，便需要一物清热消滞来平衡体内的"热气"，凉茶又成了绝佳之选。

Since ancient times, natives of Guangzhou have been foodies. It is even usual for some of them to eat seven meals a day. In Guangzhou, there is a popular saying that eating lychees accumulates much internal heat. Hence, herbal tea can be a great choice to achieve internal balance.

广州凉茶，大多并不可口，相反多少有些清苦，而这口味的"苦"却为平衡时令的"苦"而存在的。春天阴雨绵绵，湿气入体，人们困倦之苦便可靠祛湿茶之苦来平衡；至炎夏，骄阳当空，头疼脑热之苦便可靠清热毒凉茶之苦来平衡；入秋后，或有喉干咳嗽症状，奉上润肺凉茶，便可平衡津液；即便在寒冬，一杯温热的驱寒凉茶落肚，瞬间便能暖身暖心。每一位广州人眉头紧锁、愁容满面地走进凉茶铺，"咕噜咕噜"一杯凉茶下肚，那通体舒畅简直可以说是肉眼可见的。那种场景，大概与美国西部牛仔走进酒吧，喝上一杯威士忌后的神清气爽、豪气万丈相去不远吧。

Most of the herbal tea in Guangzhou tastes bitter, which aims to help relieve physical discomfort in different seasons. In spring, people will accumulate internal damp because of the rainy weather. In this case, the bitterness of the herbal tea for clearing damp does help people get rid of drowsiness. In hot summer, the bitterness of a headache can be relieved by the bitter taste of the herbal tea for clearing the heat. In

autumn, a sore throat and coughs can be cured by the lung-nourishing herbal tea. Even in cold winter, a cup of cold-dispelling herbal tea can make us warm. When people with a gloomy countenance walk into a herbal tea shop, they seem to change their moods completely after drinking a cup of herbal tea. Such a scene may be compared to that where American cowboys finish a glass of whisky in a pub, looking satisfied and energetic.

　　外地人初尝凉茶，多数只用两个字形容：难喝。然而，甘苦自知。正如北京豆汁的酸、四川火锅的辣，能让人舌头发麻发涩的"凉茶苦"，也不过是表象，喝上几口以后便觉口中生出丝丝甜意，唾液中和了茶味，似乎也就不那么苦了。再喝下去，舌上竟然又有了甘味。等到全部喝完，把杯一放，便觉满口生津，竟有唐代诗人卢仝所谓"唯觉两腋习习清风生"的韵致。

　　Non-natives would describe herbal tea as an unpleasant drink when they take the first try. However,

天时 | 第五节 | 食得健康
Right Time | Section V | Eating Healthily

one knows best what one has gone through. The sour taste of Beijing Douzhir (leftover for making green bean amylum or glass noodles), the spicy taste of Sichuan hot pot and the bitter taste of herbal tea are just initial feelings. After several sips, you will find a balance between saliva and the tea, with a slightly sweet taste. And you will taste the sweetness after trying several more times. After drinking it all, the internal heat will be relieved and you will enjoy "a satisfying cool body and mind", said Lu Tong, a poet in the Tang Dynasty.

凉茶的甘苦平衡,正是广州人"以寒凉祛湿热""据时令防疾病"之智慧的结晶吧。

The balance between sweetness and bitterness of herbal tea is created by the intelligent natives of Guangzhou, who use bitter herbal tea to clear damp and heat as well as preventing illnesses in different seasons.

"治愈系"的"广州甜"

Syrup for Cure

在 1937 年 11 月的战火中，国立西南联合大学（简称"西南联大"）是特殊岁月里的一线学术亮光。在那段艰苦岁月里，是什么抚慰着西南联大师生的心情？"广州甜"是也。就在西南联大旁边的金碧路上，一家广式糖水铺成了师生们心中的最佳"治愈系"美味之一，芝麻糊、绿豆沙、番薯糖水……在广东同学的带领下，苦中找甜大概是西南联大师生们在那段特殊时期的美食记忆，至少是著名作家、美食家汪曾祺的美食记忆。

During the war in November 1937, the National Southwest Associated University shone as the only remaining star in the academic world. In those days of hardship, what cure could be used to comfort people in the university? It was the Guangzhou syrup. A shop of Cantonese syrup, settling down on Jinbi Road near the University, provided various delicacies for teachers and students, serving as a healing place for them. Encouraged by Cantonese students, other non-native teachers and students did experience a great time of tasting delicious food in such a difficult time. At least this was the memory of Wang Zengqi, a famous writer and gourmet.

心情郁闷的时候吃点甜食,可以纾缓心情。所谓"苦中作乐",正是如此。

Sweet food can help relieve depression. Even in adversity, we could find things that cheer us up.

丝丝甜味,从来就是广州美味传承的历史脉络之一。现有文献表明,最迟晋代,甘蔗已经在广州广泛种植。南北朝已分出糖蔗和果蔗。番禺县潭洲镇的潭洲白蔗一直颇负盛名。20世纪30年代,中国内地第一家机制糖厂在广州诞生。

The sweet taste is always the vein for historical inheritance of Guangzhou delicacies. The earliest articles we have found show that, sugarcane had been widely cultivated in Guangzhou from the Jin Dynasty. In the Northern and Southern dynasties, two kinds of sugarcane could be differentiated. Tanzhou Town, situated in Panyu District, is quite famous for white cane. In the 1930s, the first machine-made sugar factory on mainland of China was set up in Guangzhou.

广州人的"擅于甜道"可见一斑。

It is obviously that the natives of Guangzhou are good at making sweet food.

糖水之所以能成为"广州限定"的美味,是因为广州糖水不像其他地方的甜汤那般单薄清淡如"糖+水",又不像西洋

天时 | 第五节 | 食得健康
Right Time | Section V | Eating Healthily

甜品那样口味剧烈，厚实饱腹。广州糖水是足料的、丰盈的，一口下去绝对是充实的——你很难找到只有一款食材的糖水；同时又是绵软的，甜味在口咽处游荡，回味无穷。

Instead of simply mixing sugar and water like that in other places or adding too much of the ingredients like that in the west, Guangzhou syrup is made up of various kinds of ingredients that help enrich the taste; and meanwhile, its soft taste lingers for long. That's why Guangzhou syrup can be unique.

在广州，跟一碗糖水打交道，就如同跟广州人交往：有料充实，却绝不生硬。

In Guangzhou, tasting a bowl of syrup is like associating with the natives of Guangzhou, substantial but not farfetching.

广州糖水以炖为主，但花样却繁多，似乎任何食材都能在糖水师傅的手中变幻出一碗糖水。鲜奶窝蛋、冰糖雪耳、番薯糖水、芝麻糊、眉豆花生番薯糖水、香芋西米露、绿豆海带沙、咖啡窝蛋奶、芒果西米露、杨枝甘露、椰汁西米露、杏仁糊、花生糊、雪耳鹌鹑蛋、木瓜桃胶……随便在街角找一家小小的甜品铺，你大概也会被密密麻麻的菜单惊到，甚至会怀疑自己是不是患上了选择困难症。

The Guangzhou syrup is usually made by stewing, but abound in variety. It seems that the cooks of the Guangzhou syrup can always make various kinds of syrup no matter what ingredients are given. Like stewed egg with milk, stewed white fungus with crystal sugar, sweet potato soup, stewed sesame-seed paste, black-sesame sweet balls, stewed sweet potato with red bean and peanut, stewed taro with sago, boiled seaweed with green bean, stewed egg with coffee and milk, chilled mango sago, chilled mango sago with pomelo, sweet sago with coconut juice, almond paste, peanut paste, stewed quail eggs with white fungus, stewed papaya with peach gum… You will be amazed by the thickly dotted menu in a common juice shop situated at the corner, and may even suspect that you have suffered the inability of making choices.

■ 木瓜炖桃胶（苏韵桦摄影）

■ 大良双皮奶（苏韵桦摄影）

  不过，还好，只要你选定了，广州糖水几乎从不会让你失望，"甜到病除"的效果，会让你进一步确信，广州糖水真的是"治愈系"美味。治愈你抑郁的心情，也未必不能治愈你的选择困难症——多来几次把店里的品种都试一遍，不就行了吗？

  However, once you make your decision, everything will be fine, as the Guangzhou syrup will never let you down. The sweetness will convince you that it is really a cure for depression and even the inability of making choices——you can cure yourself by trying the whole menu in the shop.

  而当广州张开双臂对外开放时，最先接纳外来文化的食物之一，正是最传统的糖水，如西柚、泰国芒果等，第一时间在广式糖水中浮现，其中的佼佼者自然是杨枝甘露。有着美丽名字的它，将西柚和芒果容纳其中，果肉清香、西米软糯，取悦食客的仍是熟悉的层次丰富的口感。外域风味尚且如此，邻近地区的糖水品种自然更是来者不拒，例如顺德双皮奶、姜撞奶，也早已成为广州人口福的一部分。

When Guangzhou started to open up to the outside world, it is the traditional Guangzhou syrup that become one of the first food to accomodate foreign culture. This can apparently be seen in new types of syrup made of grapefruit and Thai mangos. The outstanding one should naturally be chilled mango sago with pomelo, which does not only possess a cute name but also contain refreshing mango and pomelo with soft sago. What pleases the customers most is still its soft and rich taste. Even foreign flavor can be creatively combined into Guangzhou syrup, not to mention delicacies from neighboring places, such as double-layer milk custard and ginger milk curd from Shunde, which have become a vital part of delicious food in Guangzhou.

对于广州人来说,好的糖水不仅能满足口福,更应该有"药食同源"的保健功效。一碗小小的糖水,也讲究应季。特别是在夏季,海带绿豆沙堪称消暑神器,因为绿豆具有清热解毒的功效,而绿豆性凉,广州本地人还会放入一丝陈皮调和,这就把广州绿豆沙和其他地方的绿豆沙区别开来了。一碗糖水端上来,陈皮的香气先灌入鼻腔——这正是熟悉的家乡味道——然后才是豆沙的细腻和海带的脆爽。

For the Guangzhou natives, nice syrup does not only satisfy the appetite, but also serves as a kind of healthy food. A small bowl of syrup should be served for the right season. In particular, sweet mung bean paste with seaweed can be called an amazing cure for relieving summer heat, for mung bean is good for clearing the internal heat. But mung bean is cold in nature, so natives would add some dried tangerine or orange peel for internal balance, which helps distinguish Cantonese mung bean paste from others. When the porridge is served, the smell of dried tangerine would be felt by the nose first——the familiar sweet smell of hometown——then comes the enjoyment of soft bean paste and refreshing seaweed.

哦,对了,糖水还是粤式美味中比较少见的可以选择"冷食"的美味,与一般广府粤菜讲究"新鲜滚热辣"似有不同。不过,这轻轻的一"冷",就跟广州人与朋友交往的态度一样,"打得火热"自然不错,但有时候稍微冷却一下,说不定就会产生别样的"温暖治愈"奇味。

In various Cantonese delicacies, syrup is a rare kind of food that can be served cold, which is different from common Cantonese cuisine that features hot dishes made from fresh ingredients. However, this "cold course" is somehow similar to how Cantonese treat their friends. It is indeed delightful to be intimate all the time, but keeping a distance from each other occasionally may produce a particular warm effect that can heal our hearts.

■ 玫瑰雪莲雪耳

第二章 | Chapter II

Right Place

地利

广州依水而生，若想去哪里觅食，又无明确线索的话，可以去那些名字中带"水"的地方，他们的出品大抵不错。有水就有米，正好"水"和"米"在粤语中都有"财富"的意思，这并非巧合，岭南水乡的美味在世界上独具一派。我们未必需要艳羡日本的各种限定，在泮塘，在荔湾，在广州，可找到充满水乡情感的"限定"美味。

Guangzhou is built by the river. When you want to eat somewhere without a clear clue, you can just choose those places with the meaning of water in their names. Where there is water, there is rice. It is no coincidence that both water and rice signify wealth in Lingnan. The delicacies of Lingnan are unique. It is unnecessary to envy the various "limited supplies" of Japan. In places like Pantang of Liwan District, there are "limited supplies" of delicacies, which are representative of Guangzhou as a watery city.

第一节 | 米的幻化
Section I | **Transformations of Rice**

肠粉"奏鸣曲"
The Rice Roll "Sonata"

"早晨!牛肠加白粥啦,唔该!"

"Good morning, beef roll plus plain porridge, please."

如果你仔细统计每天早晨广州街头最常出现的语句,这一句可能是出现最多的。不要被吓到,这里说的"牛肠"并非真的牛的肠子,而是牛肉肠粉的简称。事实上,全中国没有其他地方的人,会像广州人对肠粉爱得那样"深沉"。

If you are able to calculate the most frequently sentence appearing on the streets of Guangzhou in the morning, this must be one of them. Don't panic. The beef roll has nothing to do with bovine intestines or stomachs, but a nickname (or pet name) of beef rice roll. In fact there is no other place in China where people love rice rolls deeply as much as the Cantonese do.

虽然肠粉在广州绝大部分星级酒店都能吃到，但若要真正体会其中的韵味，应该到某个街角小店去。这里说的"韵味"还真是"音韵"的"韵"。只见拉肠师傅熟练地将一个个"抽屉"从蒸箱中拉出，一勺米浆下去，再来几勺牛肉、猪肉或者反手一个鸡蛋，像极了交响乐团的指挥家，再握起"抽屉"一阵旋转轻摇，拉肠的种种内料便被安排得明明白白。此时，师傅会把"抽屉"放回蒸箱，又迅速从蒸箱里拿出另一只"抽屉"，只见"抽屉"里的米浆已是雪白剔透。四五个抽屉轮番开合，就在指尖挥洒间，生熟竟可无缝对接，犹如经过精密计算，此起彼伏地释放出动人的音符。在弥漫的水蒸气前观察一阵，你似乎就会有了错觉："抽屉"如同琴键一般起起落落，奏出的乐曲轻易就将观者的味蕾唤醒。一家正宗的肠粉档几乎变成雾气缭绕的仙境。

Though most of the star-rated hotels in Guangzhou offer rice rolls, you have to go to some corner shops if you want to experience the authentic charm and rhythm of the rice roll. A chef takes out a "drawer" skillfully, adding a scoop of rice milk, some spoons of beef, pork or an egg, just like the conductor of a symphony orchestra. Then lifting the "drawer", he spins it gently, putting all the ingredients in proper place. At this point he puts the "drawer" back into the steamer and withdraws another one. The rice milk within the latter one has turned white and clear. Four to five "drawers" take turns to be opened and closed. Just at the fingertips, the states of rice milk are seamlessly changed, as if by precise calculation. The moves are just like musical notes released one after another. Observing in front of the permeating steam, you may have the illusion that the rises and falls of the "drawers" are just like the moves of the keys on a musiccal instrument, which produce the music that easily wake up the viewers' taste buds, surrounded by flowing steam.

地利 | 第一节 | 米的幻化
Right Place | Section I | Transformations of Rice

　　其实，抽屉式肠粉只是广式肠粉的一种，广式肠粉还有两种——布拉肠和窝篮肠粉。只是这两者已较为罕见，特别是窝篮肠粉。窝篮是广东乡村用来晾晒稻米的扁平竹篮，是用来"对付"大米的工具，广州人似乎很自然地就把它拿来蒸肠粉了。用窝篮蒸肠粉可以最大限度地将肠粉的水蒸气漏掉，但要配以猛火蒸之，最好让肠粉被水蒸气吹得翘起。生活在岭南水乡的广州人当然知道"竹篮打水一场空"，但他们更关心的可能是"竹篮蒸粉一阵竹香"，因此窝篮肠粉是在广州西关才能找到的"广州限定"。

　　In fact, drawer-style rice roll is just one type of Cantonese rice rolls. There are two others–cloth and bamboo basket ones, but they are rarely seen these days, especially the last one. Bamboo baskets are used for drying rice in Guangdong villages. As they are also tools for cooking rice, the Cantonese naturally use them to steam rice rolls. Bamboo baskets can maximally drain excessive steam, so the rice rolls can curl up when steamed with high fire. Of course, the Cantonese know that "it is futile to draw water with a bamboo basket", but they also know that it adds to the flavor to steam rice rolls with bamboo baskets. The bamboo rice roll is therefore one of the "limited supplies" of Guangzhou that can only be found in Xiguan.

不过,无论是采用哪种制作方法,一碟正宗的肠粉,首先看的是粉皮。观之薄如蝉翼、晶莹剔透,啖之细腻爽滑、不失韧性,即为上品。特别是布拉肠,在米浆倒出来后,师傅可以直接"下手",运用熟练的手法均匀地将米浆往"四面八方"推开,而不像"抽屉式"的那样只能靠"摇"。若是布拉肠师傅的手势足够出色,便可保证米浆受热均匀,拉出的肠粉也能够更薄——只有够薄,才可以让食客在爽滑弹牙与绵软回味中感受到恰到好处的平衡。

However, whatever production method is adopted, the most important part of a plate of rice roll of high quality is the wrapping. If it looks as thin and crystal clear as cicada wings and tastes smooth but elastic, it is of the highest grade. It is especially true for cloth rice rolls, because after the rice milk is poured out, the chef can use his hand directly and spread it in all directions instead of shaking it like drawer-style rice rolls. If the chefs are skilled enough, the rice milk will be evenly heated and the rice roll will be even thinner. Only when it is thin enough can it strike a balance between smoothness and softness.

地利 | 第一节 | 米的幻化
Right Place | Section I | Transformations of Rice

　　米浆本无味，要靠酱油提鲜。据说许多广州肠粉师傅什么都可以教，但配的酱油总坚持"秘制"。肠粉用的酱油不能只是一"咸"了之，因为只"咸"就不"鲜"了，有些许甜味才可以烘托出肠粉内料的鲜味。

　　The rice milk is tasteless, so soy sauce is needed to "lift" it. It is said that many rice roll makers in Guangzhou are willing to teach you anything but the soy sauce. They insist on keeping the soy sauce a secret. It cannot be just salty; otherwise it is not savory enough. A little sweetness can enhance the freshness of the rice rolls.

　　虽说肠粉除了在广州都可以吃到，但愿意在肠粉上玩点新花样的恐怕就真的只有粤菜师傅了。例如，凉瓜牛肉肠，直接将经典搭配的凉瓜牛肉加进肠粉里，给牛肉肠增加一丝清凉；烧鹅肠粉，用烧鹅汁代替酱油，再来点芝麻和葱花，实力演绎"金秋玉露"。

　　Even if you can have rice rolls in other places, you might find only the Cantonese chefs who make rice rolls are willing for innovation. For example, bitter melon and beef are added to the rice rolls to make them taste cool; roast goose rice rolls, where the chefs replace soy sauce with roast goose sauce, and add some sesames and shredded shallots, making it more colorful.

肠粉通常都是"啖啖肉"，鱼片肠似乎已经是肠粉和鱼肉结合的"极限"。但广州人不断追求，创制出新的爆款鳕鱼肠粉，它用幼嫩鲜美的加拿大香煎银鳕鱼做馅，再用传统西关布拉肠做皮，兼有传统与创新。就连多骨的鲫鱼，广州人也敢尝试给鲫鱼和拉肠做起"媒人"，促成一对"好鸳鸯"的，大概绝不会有第二个了。师傅先从鱼鳃处开刀，沿着鱼脊将鱼肉和鱼背分离开，切成若干块 5 厘米长的鱼段。最重要的步骤是起骨，用刀按 U 字形将鱼背肉的骨刺"起"出来，部分比较难"起"的细刺需专门将其钳出，而鱼腹肉上的骨刺比较粗，只需用刀斜切一次就起完。骨刺"起"完后，剩下的肉切片即可。一条鲫鱼的"起肉率"最多也就四成。将鲫鱼肉加上均安大头菜等埋入肠粉内蒸熟，淋上葱花和酱油便大功告成。

Usually rice rolls go together with meat and fish fillet seem to be the limit of combinations. However, the natives of Guangzhou keep on innovating. The new favorite is called sautéed Canadian silver pout rice rolls, featuring both tradition and innovation. Even the thorny crucians are put into rice rolls. The natives of Guangzhou are like go-betweens who have successfully matched the two. The chef cuts the fish from the gills, separating the fillets from the back along the ridge and cutting it into pieces of 5cm long. The most important step is boning, removing the bones from the back in U-shapes. Some finer bones need to be clamped out, while the bones on the belly are large enough to be removed all at once. After boning, the remaining fish is cut into fillets. The "fillet rate" of a fish is at most 40%. The fish fillets and Jun'an preserved cabbage are buried in the rice rolls and steamed together. Once done, add some shredded shallots and soy sauce.

地利 | 第一节 | 米的幻化
Right Place | Section I | Transformations of Rice

由此可见，肠粉几乎可以包裹任何食材，猪肉、牛肉、鱼片、虾仁自不待言，鸡蛋、油条、菌菇、时蔬等也可"一包了之"，难怪有人形容肠粉仿如广州人一般，拥有开放的怀抱、温和的性格，从不拒绝新的搭档。

As can be seen, rice rolls can wrap up almost everything, like pork, beef, fish, shrimps and even eggs, deep-fried dough, fungi and vegetables. It is no wonder that some people say rice rolls are just like the Cantonese, with open minds and gentle characters, never rejecting any new partners.

干炒牛河(苏韵桦摄影)

非遗一箸沙河粉

The Intangible Cultural Heritage-Shahe Rice Noodles

广州有个地方叫沙河镇，镇上有一种岭南名小吃叫"沙河粉"。它洁白薄韧，如豆蔻少女之肌肤；米香浓郁，吃得人啧啧称赞。

In the northern suburb of Guangzhou, there is a Town called Shahe , in which there is a famous snack called Shahe Rice Noodles. It is white, thin and elastic, just like the skin of a girl. It has the strong aroma of rice.

沙河粉从何而来？
Where did Shahe Rice Noodles come from?

时间定格在清末。沙河镇上有家小食店，店主名樊阿香。一日，门前来了一位衣衫褴褛的长者，好心的店家夫妇给老人家送上热粥。谁知老人家后来天天到店里喝粥，夫妇俩也未曾嫌弃。有一天，阿香生病不思饮食。老人得知后，从白云山上取来泉水浸米，磨成米浆，将米浆舀入窝篮上薄薄摊开来蒸，米浆熟后成为粉皮，老人将粉皮切条加油盐香葱后送予阿香。阿香一试，胃口大开，连吃几份。后来阿香方知老人原是宫中御厨，逃难至此。因身处沙河，老人给粉起名叫"沙河粉"。

Dating back to the late Qing Dynasty, there was a small snack shop in Shahe Town. Its owner was called Fan Axiang. One day, there came an old man in rags. The kind-hearted couple treated him with hot porridge. But then he came to the shop every day for porridge. The couple never minded. Later, Axiang was sick and had no appetite. After the old man knew it, he fetched some water from the Baiyun Mountain to soak the rice and ground it into rice milk, which was then spread thinly on a bamboo basket and steamed. The rice milk turned into wrappings. The old man cut the wrappings into slices, added some oil, salt and shallots and served them to Axiang. After trying them, Axiang had a good appetite, and ate several servings. Later Axiang came to know that the old man was originally an imperial chef who fled from calamities. Because the shop was located in Shahe, the old man named the food Shahe Rice Noodles.

沙河粉命运坎坷，曾经因制作工序繁复而渐被放弃，连原本做沙河粉非常拿手的沙河大饭店都因经营不善而倒闭。后来，广州市政府立项将沙河粉传统制作技艺交由开平人区又生创办的"沙河粉村"传承。2002年，区又生专程向沙河粉传统制作技艺第四代传人陈丽珍及伍谷学习制作工艺，经过一年半载的磨炼，他做出来的沙河粉薄如纸、通透明亮，透过粉皮甚至可以看到字。在区又生的努力下，"沙河粉传统制作技艺"成为广东省非物质文化遗产之一。沙河粉的制作最重要的是用水、选米、磨浆、蒸粉四点，按照旧时做法，全手工制作。将开平山地生态米浸数个钟头后加帽峰山泉水磨成浆，舀一椰勺的量，均匀倒在窝篮上，放在滚沸的水上蒸，成形即揭开。

Shahe Rice Noodles have suffered a frustrating fate. Because of its complicated production process, it was once abandoned. Even Shahe Restaurant, which was good at making Shahe Rice Noodles, went bankrupt due to poor management. Later,

地利 | 第一节 | 米的幻化
Right Place | Section I | Transformations of Rice

the Guangzhou Municipal Government approved the project to transfer the manufacturing techniques of Shahe Rice Noodles to "Shahe Rice Noodle Village" established by Ou Yousheng, a native of Kaiping. In 2002, Ou learned the techniques from the fourth-generation successors Chen Lizhen and Wu Gu. After painstaking practices for some time, he made the rice noodles as thin as paper and transparent. Words can even be seen through the rice noodles. With Ou's hard work, Traditional Shahe Rice Noodle Craftsmanship has been listed as one of Intangible Cultural Heritages of Guangzhou. The four most important factors that influence the taste of Shahe Rice Noodles are water, rice, grinding and steaming. It is completely handmade, following the traditional practice. Ecological rice grown in the hilly fields of Kaiping is soaked for hours and ground into milk with water from the Maofeng Mountain. The amount of one coconut spoon is spread evenly on the bamboo basket and steamed over boiling water. Once transformed, the wrappings are removed.

"沙河粉村"位于白云山下云台花园边，港澳朋友来广州少不了到此"打卡"。一来可以试一试传说中的沙河粉，毕竟大家都是热爱干炒牛河的人。"牛河"正是牛肉河粉的简称，而"河粉"是"沙河粉"的简称。二来可以顺路上白云山呼吸新鲜空气。中国香港众多知名人士甚至多次前来，只求过过嘴瘾。曾有诗

■ 干炒牛河（苏韵桦摄影）

言："非遗一箸沙河粉，情倾亿万南粤人。"其实何止南粤，整个粤港澳大湾区多少老饕为之而来。一箸沙河粉，如同绳索将粤港澳三地紧紧牵连，汇集于舌尖。

　　Shahe Rice Noodle Village is located by Yuntai Garden at the foot of the Baiyun Mountain. It is a must-visit place for visitors from Hong Kong and Macao in China. On the one hand, they can try the legendary Shahe Rice Noodles, because a lot of people are crazy about stir-fried beef rice noodles. On the other hand, they can enjoy the fresh air of the Baiyun Mountain. Many celebrities from Hong Kong of China have paid special visits to this place village to enjoy the delicacy. The verse says "As the intangible cultural heritage, Shahe Rice Noodles is the love of hundreds of millions of Cantonese." It is not just Cantonese but many foodies in the Guangdong-Hong Kong-Macao Greater Bay Area that have come for it. Shahe Rice Noodles are just like a rope, closely connecting the three places of Guangdong, Hong Kong and Macau.

地利 | 第一节 | 米的幻化
Right Place | Section I | Transformations of Rice

沙河粉最经典的演绎，莫过于深入人心、纵横粤港澳乃至东南亚地区的干炒牛河。河粉油润亮泽，牛肉滑嫩焦香，镬气加持，热气腾腾，无论何时都是慰藉南粤人辘辘饥肠的最佳美味。沙河粉的路子远不止干炒牛河，它还能走得更远，做出咖喱炒河、五色河粉、鱼蛋汤河、榨菜肉丝河，甚至做成馎饦、西式比萨。几十种沙河粉的做法，将粤人创新永不止步的精神浓缩在餐桌上，浓缩在这一碟有着浓浓岭南风味的沙河粉上。

The most classic interpretation of Shahe Rice Noodles is the stir-fried beef one which has won the hearts of people across Guangdong, Hong Kong and Macau in china and even in Southeast Asia. The noodles are bright and glittering, and the beef smooth, tender and aromatic. Served hot, it is always a delicacy for Cantonese. Shahe Rice Noodles can go even further with more varieties like stir-fried Shahe Rice Noodles with curry, five-colored Shahe Rice Noodles, fish ball Shahe Rice Noodles with soup, Shahe Rice Noodles with shredded meat and preserved vegetables, or they can be made into thin cakes and even western style pizzas. The several dozen ways of cooking Shahe Rice Noodles have demonstrated the spirit of the Cantonese who are striving forward.

■ 制作水菱角（王维宣摄影）

"瀨"出米的"味"力

"Flowing" Tastes

欲采新菱趁晚风，塘西采遍又塘东。满船载得胭脂角，不爱深红爱浅红。

——广州西关民歌《采菱》

Picking water caltrops in the evening wind, first west of the pond then east of it, loading the boat full, lighter ones better than darker ones.

—— Guangzhou Xiguan folk song
Picking Water Caltrops

"泮塘"之地，居于广州西部，以广州之大，泥土肥沃却几乎无过此地。追溯远古，此地为一片池沼，只因沧海桑田，泥沙冲积日久，珠江一侧的汪洋也成了池塘洼地，因此有了"半塘"之名。广州人"以水为财"，于是干脆在"半塘"的"半"字边上再加"三点水"，终成"泮塘"。

"Pantang" is located in the west of Guangzhou. It is the most fertile place in Guangzhou, because in ancient times, it was a pond. The alluvial sediments over time, turned part of the pond facing the Pearl River into marshes, hence the name "Bantang" (half pond). Since natives of Guangzhou regard water as wealth, they add water to the Chinese character meaning "half" in the name , making it "Pantang".

清香飘逸的池塘洼地之间，菱角带着水乡秋天的气息出水。巧手的西关姑娘们挽起菱盘，轻巧地将菱角从泥中拈起，又轻轻放下，生怕破坏了其"双角"。以菱角为重要组成部分的"泮塘五秀"（指菱角、慈姑、莲藕、茭笋、马蹄）似乎在提醒世人：我西关一地，人杰地灵，能称"秀"的可不止西关小姐。

Water caltrops ripen in autumn. The skilled Xiguan girls carefully lift them from the mud and place them lightly on the plates, for fear of destroying their "double horns". The water caltrop is one of the "five beauties" of Pantang in Xiguan (including the water caltrop, arrowhead, lotus root, wild rice stem and water chestnut), as if reminding the world that Xiguan is a place blessed with various talents, not just Xiguan ladies.

要品尝新鲜的菱角，自然要来泮塘。菱角之鲜美，正在于"泮塘限定"，同时也在于"中秋限定"。中秋月圆之时，鲜甜的菱角曾穿梭于西关的骑楼之间，广州老饕们的脚步踩遍了西关大街的青石板路，赏月赏味的闲适气氛在珠江边流转。

The best place to taste fresh water caltrops is Xiguan, the place of origin. In the past, it was precious because it was of "limited supply in Xiguan" and "limited supply during the Mid-Autumn Festival". When the moon was at its fullest, the sweet fragrance of water caltrops was permeating along the arcades. Foodies in Guangzhou trod on every stone-laid road of Xiguan, enjoying the moon and the delicious water caltrops by the Pearl River.

然而，深秋一过，菱角何处寻？直到清末，王玉明正式登场。王玉明婆婆住在带河路（留意到吗？又是一个带"水"的地名），即便在没有菱角的季节，大户人家也总是有大米的。王婆婆不一定熟耍锄头，但筷子功绝对一流，因为她居然想到用一双筷子"濑"出新鲜菱角的形状。所谓"濑"的手法，是这样一套魔术般的动作：筷子在米浆中一按一提一拨，迅疾地将米浆拨入热水中，米浆竟乖乖地在水中"长出双角"，最后展开如菱角的形状。一按一提一拨，看似信手拈来，实则需要"稳、准、狠"，分毫不差。

However, when the late autumn was over, the water caltrop was nowhere to be found. By the end of the Qing Dynasty, Wang Yuming appeared. Granny Wang lived on Daihe Road. Even when there was no water caltrop, there was rice for rich families. Granny Wang might not know how to use the hoe, but she was first-class at using chopsticks. She even thought of creating the shape of water caltrops by flowing the rice paste. The actions are quite magical, pressing and then lifting the chopsticks from the rice paste and quickly placing them in boiling water. Then the rice paste is changed into a "water caltrop". Though the actions seem quite simple, they actually need to be "steady, accurate and decisive".

大米和水的搭配亦是关键。"濑"之前，粘米粉要用80℃左右的温热水兑开；"濑"之时，水温太高粘米粉会散开，水温太低粘米粉又粘在一起，皆不成角状。这就要求"濑"菱角时的节奏需如演奏交响乐般精准，水温高时要拼手速，水温低时需轻慢些；"濑"之后，当水菱角在滚水中浮起，就可以捞出来"过冷河"，一热一冷之后，水菱角就变得十分弹牙。

The combination of rice and water is also the key. Before "flowing", rice flour needs to be mixed with water of about 80℃. If it is too hot, the rice flour will be scattered;

if too cold, the rice flour will be too sticky. So making rice water caltrops should be as accurate as playing a symphony. When the temperature is high, one has to be quick; when it is colder, one has to be slower and gentler. When the rice water caltrops flow up, they can be scooped up and placed in cold water. With the heat and cold, the rice water caltrops become very elastic.

如此"濑"功，竟影响了广州人的婚嫁大事。以前的西关小姐在出嫁前，不比自拍之浮夸，也不比婚纱之豪华，就喜欢相互切磋"筷子功"。小姐闺中，飞舞的不只是针线，大概还有筷子。

Such "flowing" skills have actually influenced the marriage customs in Guangzhou. In the past, the Xiguan ladies would exchange their skills of using chopsticks as well as needle work before their weddings instead of boasting their luxurious wedding gowns.

水菱角的原料是大米，浓浓米香自不待言，但广州人的味觉可不能被如此敷衍了事，老火猪骨汤必不可少，再加上花生、冬菇、虾米、腊味一同配制，入口方能层次丰富。柔顺嫩滑的那一抹白，还得拌点炒香的大头菜、猪油渣、鸡蛋丝、辣椒圈或叉烧丝，于是碗里瞬间灵动了起来。

The raw material of rice water caltrops is rice. Though it has the strong aroma of rice, the taste buds of Guangzhou natives demand more. So pig bone soup is a must, together with peanuts, dried mushrooms, shrimps, and preserved meat to enrich the taste. The white and smooth rice water caltrops need to be mixed with some

地利 | 第一节 | 米的幻化
Right Place | Section I | Transformations of Rice

preserved Chinese cabbage, cracklings, egg slices, peppers or roast pork slices to make them more delicious.

广州人吃水菱角可不仅仅是"假装在吃菱角"那么简单，在复制天然菱角之余，亦可见泮塘人对水乡独一无二的浓厚感情，终使广州泮塘成为食客觅食水菱角的不二之选。正是泮塘人对大米和菱角两种天然食材的特殊情感，才酿出水菱角的独特风味。

When eating rice water caltrops, Guangzhou natives are not simply eating the caltrops. It has also reflected the profound feelings of Pantang people towards their riverside hometown, and has made Pantang the best place to enjoy rice water caltrops. It is their special feelings towards rice and water caltrops that have created the unique feature of rice water caltrops.

毕竟制作濑水菱角要求技艺极高，能做出来的店在泮塘乃至整个广州也不过一两家，若愿略微降低要求，亦可通过手捧一碗濑粉，体会广州人对大米的喜爱。同样是"濑"的手法，濑粉就简单得多。如果说濑水菱角是大家闺秀，濑粉定然就是邻家女孩。米浆从一种扎有众多小孔的器皿慢慢流出，细长、雪白的粉条如瀑布般落入锅中，待其出锅时已是弹牙爽滑，其温润清新，岂亦非迷人？从泮塘出发，濑粉早已是岭南名食，东莞烧鹅濑、佛山高明濑粉闻名海外。粤港澳大湾区东西两岸，早已通过一碗濑粉紧密相连。

Making rice water caltrops demand exquisite skills. In Pantang and even the whole Guangzhou, only one or two restaurants can make them. If customers are willing to lower their demand, they can enjoy a bowl of flowing rice noodles and experience the love for rice by Guangzhou natives. The method of making flowing rice noodles is much simpler. If the rice water caltrops is a lady from a celebrated family, the rice noodles is a neighborhood girl. The rice milk slowly flows through a container with many holes, so the thin and white rice noodles fall into the pot like mini waterfalls. When they are scooped out again, they become smooth and elastic. Aren't they also amazing? Flowing rice noodles from PanTang have become a famous food of Lingnan. Dongguan roast goose flowing rice noodles and Foshan Gaoming flowing rice noodles have been famous at home and abroad. Even before the Hong Kong-Zhuhai-Macau Bridge was built, people had already been linked by flowing rice noodles.

不过，千万不要以为濑粉就无须考究。之所以在"濑"的时候能做到随手可得，完全是因为将最关键的功夫用在制作米浆上。制作米浆适用的米，无论是用糙米还是精米，都要用冷水浸足 3 个小时，后研磨成粉。磨好的米粉细腻雪白，再自然晒干。将生熟粉一起开浆，把粉烫熟，通过流质与温度来判断粉浆的熟度，最后加入煲好放凉的米，才制成合格的米浆。

■ 烧鹅濑粉（卢政摄影）

However, please don't have the false impression that flowing rice noodles are not exquisite. The process seems so simple only because all the efforts have been given to making the rice milk. No matter it is brown rice or polished rice, the rice has to be soaked in cold water for three long hours before

地利 | 第一节 | 米的幻化
Right Place | Section I | Transformations of Rice

being ground into flour, which is white and fine. Then it is dried in the sun. The raw and cooked flour is mixed together to make the rice milk. The fluidity and the temperature determine the readiness of the rice milk. Finally some rice that has been cooked and cooled is added till qualified rice milk is ready.

种种繁复的步骤，正应了那句话：你必须非常努力，才能看起来毫不费力。常听人评价广州人为"既喜享受闲适，又能勤奋营生"，这两者矛盾吗？来泮塘吃一碗水菱角或濑粉吧，你的味蕾自然会告诉你答案。

The various complicated steps are summarised into the buzzword : you must try very hard to look effortless. The natives of Guangzhou are said to "love enjoyment and leisure but work hard as well". Are they contradictory? Come to have a bowl of flowing rice water caltrops or rice noodles, and your taste buds will surely tell you the answer.

■ 水菱角（王维宣摄影）

■ 艇仔粥（苏韵桦摄影）

淡薄粥中滋味长

Long- Lasting Flavor in Insipidity of Porridge

凉风送爽，荔湾晚唱。

在水中央，丽影一双。

一者艇仔，二者艇仔粥。

In the cool breeze, singing comes from the Lychee Bay at dusk.

In the middle of water exist a pair of shadows.

The boat and the Sampan Congee.

荔枝湾上，一艘小艇从远处轻摇来，艇上传来疍家咸水歌声般的叫卖声："艇仔粥，食艇仔粥咯……"粥香随着叫卖声和点点渔灯渐近。

On the Lychee Bay, a small fishing boat is drifting from afar. The Tankas, the boat dwellers, are calling out on the boat, "Sampan Congee, come over to eat…" Along with the calling, the fragrance of porridge and the fishing lights are approaching.

这景象，始于明末清初。当时，广州荔枝湾已聚集了一批富商大贾，亦有不少文人雅士在此"游船河"。及至清末民初，十三行正鼎盛，各路商船云集珠江口,这里成为商家的休闲所、渔家的生意场——就地打捞小虾、小鱼、小螺,再加上熬好的粥底,便可简单快捷地捧出一碗艇仔粥卖给商人们。一时间,小艇、粤曲、粥香齐飞,好一派商都兴盛景象。而艇仔粥香,随之延绵至今。

This scene can date back to the late Ming and early Qing Dynasties. At that time, there had been a bunch of rich merchants gathering at Lychee Bay in Guangzhou and many scholars sightseeing here by boat. In the late Qing Dynasty and some years after 1912, the Thirteen Hongs was booming as merchant ships of all kinds gathered at the Pearl River estuary, a leisure place for merchants and a business place of fishermen, who caught fish, shrimps and snails on the spot and added them into a well-cooked congee base. Easily and quickly there came a bowl of Sampan Congee which could be sold to the merchants. The fishing boats, the sound of Cantonese opera and the smell of porridge came together to present what a flourishing city Guangzhou was, with the fragrance of Sampan Congee lasting to this day.

河水本索然,小艇添活力;粥水本清淡,足料艳夺目。下足料的艇仔粥,红(花生)、黄(蛋丝、炸面)、绿(葱花)、琥珀(海蜇)等色散落四方,均匀有致,相互映衬,在夜色中更显多姿多彩。

As small fishing boats instill vigor into the boring rivers, bountiful ingredients make the insipid porridge attractive. Sampan Congee consists of different colors, such as red (with peanuts), yellow (with egg slices and fried noodles), green (with chopped spring onions) and amber (with jellyfish). With colors evenly scattered and contrasting each other, Sampan Congee becomes more appealing at night.

不过,在多彩粥料之余,各位食客也不要小看了那隐于其中的"白"。讲究的大米及其制法,是一碗艇仔粥成功的关键。粥水讲求浓稠适中,所以粥底要黏稠但不能过干。一碗艇仔粥里,水太多米太少就成了"米汤",米太多而水太少就成了"稀饭",都入不了挑剔的广州人的眼。广州人喜爱顺滑爽口的粥底,粥里的米不仅要"开花",还要做到绵滑黏稠如"水米交融",不少店家的粥里面甚至找不到米粒。

Apart from the colorful ingredients, eaters should not overlook the "white" color of the porridge. The key to a bowl of tasty Sampan Congee is the rice and the way of cooking. The porridge should be moderately dense and not too dry. Sampan Congee with too much water and too little rice becomes "rice soup" or "rice gruel" vice versa, either of which would not win over finicky Guangzhou natives. A congee base that satisfies their taste should be smooth and refreshing, which requires the rice not only to "bloom" but also to be dense and smooth to the extent that the rice is well integrated with the water. In the porridge served by some shops, grains of rice can hardly be found.

按照一般粥品的做法,厨师只要用经水浸泡足够时间的大米,便可慢火煲粥了,但是对于爱好吃的广州人来说,这是不够的,吸饱满水的大米,还要放入冰箱急冻,使之更容易达到"水米交融"的效果。而且,做正宗艇仔粥的米须选用传统的岭南长型稻米,因为其富含直链淀粉,大米易于水溶,更易达到黏稠的口感;倘若选择东北米(短型稻米),大米就很难"开花"了。

In general, a chef can cook rice that has been soaked long enough with gentle heat to make porridge. However, it's far from what the picky Guangzhou natives want. After a soak, the rice should be frozen so that the rice can be easily integrated with water. The traditional long rice in Lingnan is a good choice to make Sampan Congee because it is rich in amylose which can facilitate the disintegration of rice and make the porridge denser. However, the short rice from Northeast China is difficult to disintegrate and "bloom" while cooking.

艇仔粥旧时又有"渔获粥"的别称。渔获是粥料的灵魂,粥料可以多达8~10种。鱼片、河虾、猪皮、鱿鱼丝、叉烧丝、烧鸭丝、蛋丝、炸面、花生、螺肉……渔家自己是不怎么讲究贵价食材的,渔获里剩下"虾毛仔"等边角料,都可以用来做艇仔粥,偶尔出现的海蜇、小螃蟹则可能成为自家粥里的"小确幸"。

■ 西关艇仔粥

地利 | 第一节 | 米的幻化
Right Place | Section I | Transformations of Rice

In the old days, Sampan Congee was known as "Yuhuo porridge". Yuhuo, namely fishing harvest, is crucial to the porridge ingredients ranging from 8 to10 kinds, for example fillets, river shrimps, pig skin, squid slices, roast pork and duck slices, egg slices, fried noodles, peanuts, and snails. Fishers are not finicky about expensive ingredients. Whatever harvested in fishing can be added to the Sampan Congee served for others, while the occasionally caught jellyfish and small crabs may become the "small fortune" in porridge made for their own.

多彩的艇仔粥配料，关键不在于昂贵，而在于新鲜，无论是鱼片还是猪肚、猪皮还是葱花，只要够新鲜，就足以征服广州人的味蕾。也有小艇商家为做出新意，用蚝豉和干鱿鱼来调味，鲜美得让食客"眼前一亮"。

Freshness other than the price is essential to the various ingredients of Sampan Congee. Fillets, pig stomachs, pig skin or shredded green onions can all satisfy the taste of Guangzhou natives as long as they are fresh. There are also fishers who add dried oysters and squid to the porridge to make it more delicious.

轻轻舀起，米粥挂羹，滑过口舌，慢流入胃……你或许便对这粥无限留恋：这鲜美，这香甜，请你们走得慢一些，再慢一些吧。

Gently scoop up the porridge which slides over the tongue and then slowly flows into the stomach... You may be infatuated with this delicious and sweet porridge, praying that it would go slowly.

　　以往的水上艇仔粥店家，在照顾食客方面可谓十分体贴。商贾谈天说地，及至深夜，需要能饱腹的粥，因此当年的渔家们煮粥时都会特意将粥煮得比较稠，好入口之余，也为长时间在船上的商人们的胃着想，会比较好消化。另外，以往不少店家会用明火一直煲着艇仔粥直至上桌，使整碗艇仔粥一直保持着暖胃暖心的温度。当然，如今的艇仔粥早已"上岸"，你若漫步西关一带，偶遇粥水小铺，又恰见食客盈门，不必做他想，直接入内，叫上一碗艇仔粥。即便排队者众多，当粥水入口，自然会深感等待极其值得。

　　Previously Sampan Congee vendors served the eaters considerately. At that time, merchants often talked over things till midnight and needed porridge that could fill their stomachs. Therefore, the fishers would cook the porridge dense not only to whet their appetite but also to be easily digested by merchants who had stayed long on a boat. And quite a few fishers would keep heating the porridge with fire until the Sampan Congee had to be served, so as to keep the porridge at a temperature that brought warmth to the eaters' bellies and hearts. Nowadays, Sampan Congee is no longer made on boats.

地利 | 第一节 | 米的幻化
Right Place | Section I | Transformations of Rice

If you linger in Xiguan and come across a porridge shop which is swarming with eaters, don't hesitate to enter the shop and order a bowl of Sampan Congee. Despite the long queues, when the porridge is in your mouth, you will realize that the waiting is extremely worthwhile.

广府人喜好粥水，而牛肉粥、瘦肉粥、鱼片粥可以在其他地方找到，唯独这碗艇仔粥，它可算得上是"广府限定"。也许你能复制顺滑的粥水，但你能在其他地方找到如此新鲜的小鱼、小虾吗？即便能找到以上种种鲜物，又能在哪儿复刻河中小艇穿梭、湾畔树影婆娑的都市桃源呢？

Cantonese are fond of porridge. Beef porridge, pork porridge and fish porridge can be found elsewhere while Sampan Congee is only served in Guangdong. You may cook such smooth porridge but can you get such fresh fish and shrimps in other places? Even if you can get all of them, where else can you see the drifting boats on the rivers and the shadows of trees at the bay?

明代诗人张方贤写道："莫言淡薄少滋味，淡薄之中滋味长。"这悠长的滋味大概不仅仅是粥的美味吧。广府人可以坐着千吨舰艇出洋、搭乘万吨巨轮离乡，但其心中的乡愁不过一艘小艇，当然还少不了艇上唱咸水歌的姑娘和一碗热粥。

　　As Zhang Fangxian, a poet in the Ming Dynasty said in his "Poem on Porridge", "Don't say the porridge is insipid with lack of flavor. Instead, it has a long-lasting flavor in its insipidity." Such a profound feeling is beyond the taste of porridge. Cantonese can go overseas on a huge ship, but what they really miss is simply a small boat in which girls may be singing Tanka songs and a bowl of hot porridge is served.

第二节 | 小吃不小
Section II | More than Just Snacks

云吞面，是面还是云吞？
Wonton Noodles? Noodles or Wonton?

每每聚餐见有人点云吞面，我都向他们提出"云吞面是面还是云吞"这个问题。通常得到的回应是一句："你好无聊哦。"对广州美食文化有研究的人会告诉我：管它是面还是云吞，最重要的是刚煮好的时候，面条要覆盖在云吞之上，避免浸在汤里，这样面条才不会吸收过多面汤，口感也不至太软。

Whenever seeing someone ordering wonton noodles when we dine together, I would ask the above question. The usual response would be "Hah, interesting!" But if someone knows a little bit about Cantonese cuisine, he or she may tell me: no matter it is noodles or wonton, the most important thing is that the noodles should be placed on the wonton when it's ready to serve, so that the noodles would not absorb too much soup to become too soft.

不过，对于这个不是问题的问题，云吞有话要说——

However, for this problem that cannot be counted as a problem, wonton has something to say.

一碗云吞面的"主角"当然是云吞。君不知，早年的云吞面连规格品类都是按云吞的形状来限定的。20世纪30年代，云吞面开始在广州流行，但并不是在固定档位售卖，而是"走鬼"沿街叫卖。于是就有了这样一幅画面：骑楼之上，想喝下午茶的西关小姐推开窗户，唤一声楼下的云吞面师傅；骑楼之下，师傅则将煮好的云吞面放在一个箩筐里用绳索吊上楼，西关小姐伸出纤纤玉手将钱放在箩筐里，又重新往楼下吊。

The "protagonist" of wonton noodles is of course wonton. You may not know that, in the early days, the size of a bowl of wonton noodles was decided by the shape of the wonton. Around the 1930s, wonton noodles began to gain popularity in Guangzhou. But at that time, wonton noodles were sold by wandering vendors rather than in fixed stalls. Thus, there were once scenes like this: on the arcade, Xiguan ladies, who wanted to eat afternoon tea, would open the window elegantly and order wonton noodles from the wandering vendors; under the arcade, the vendors would lift the wonton noodles upstairs in a basket, and then the ladies would lower the basket back to the venders with money.

云吞面根据规格大小，分别叫作"大用"或者"细用"。后来，当云吞面从廉价小吃渐登大雅之堂时，文人们便觉得这"用"字太庸，而改用"蓉"。加之煮熟的云吞仿如盛放的芙蓉花，"蓉"的美称就此传开。

■ 虾子面

Wonton noodles were once called "Da Yong" (large size) or "Xi Yong" (small size) according to the size. Later, when wonton noodles grew increasingly elegant from cheap snacks, the literati felt that "Yong" was too mediocre and replaced it with "Rong" (hibiscus). What's more, cooked wontons look just like hibiscus flowers in full bloom, so the name "Rong" spread.

"细蓉"一般有4粒云吞，6粒云吞的是"中蓉"，而8粒云吞称为"大蓉"。这一"蓉"字，竟将广州人的饮食美学，以一种奇妙的方式与花城的"花"产生了某种微妙联系，巧合哉？

"Xi Rong" (small size) generally has 4 wontons; "Zhong Rong" (medium size), 6; and "Da Rong" (large size), 8. Isn't it wonderful that the single character "Rong" should produce a subtle connection between the food aesthetics of Guangzhou natives and the "flowers" from the "City of Flowers", the nickname of Guangzhou?

云吞皮展开如芙蓉花开，入口丝滑，此特性主要源自用料和手工，缺一不可。云吞皮选用精面粉、鲜鸭蛋作为原料，纯手工和面，1斤面粉，0.5斤鲜鸭蛋、4钱（1钱=5克）碱水搓至起筋，然后用打面机或人工碾压成薄如纸的云吞皮，每斤面大约可打出260张云吞皮，可见其薄之极。只有达到这个程度，才能够创造出爽滑的口感。

Spreading wonton wrappings look like hibiscus flowers, and have a smooth texture. This feature is mainly attributable to ingredients and handwork, both integral components. Combine every 500 grams of fine flour with 250 grams of fresh duck eggs and 20 grams of alkaline water, and then knead with hands until gluten is extracted. Then, knead the dough by hand or put it into a pressing machine to make wonton wrappings. Every 500 grams of dough makes about 260 pieces of wonton wrappings, which shows how thin each is. Only in this way can the smooth taste be created.

云吞馅料则采用新鲜猪肉，八成瘦肉、二成肥肉（也有师傅爱用"三分肥七分瘦"），把肥肉与瘦肉分离，肥肉去掉泡腩，瘦肉剔筋，手剁成馅。拌上碾碎并经过油炸的大地鱼肉和除壳的鲜虾仁、鲜鸭蛋黄（1斤肉馅对应加4个蛋黄），再调上味粉、熟盐、糖制成。馅料做到干湿适宜。手剁猪肉层次分明，使猪油得以圆满保存，并丝丝渗入馅料之中，才有了入口时全方位的口感。

As for the fillings of wanton, fresh pork is a must. First, separate the fat and lean parts of the pork, and then remove the bubbles from the fat and the membrane from the lean part. Then, mix the lean and fat part with the ratio of 8:2 (or 7:3 as some cooks prefer), also with smashed deep-fried flounder and shelled fresh shrimps, fresh duck egg yolks (four egg yolks for every 500 grams of fillings), along with condiments, cooked salt and sugar. The fillings must be of suitable moistures. It is the hand-chopped pork that keeps the lard inside which can then infiltrate the fillings, making the taste so full.

面条也有话要说——
Noodles also have something to say.

"细蓉"一名来自芙蓉花没错，但这里的"芙蓉"可不是用来形容云吞，而是形容面条的。说法之一是，之所以称云吞面为"蓉"，是从"芙蓉面"衍生而来。"芙蓉"是形容外貌美丽的女子，故在中国古代文学中可比作靓面。说法之二是，"芙蓉面"源自唐代诗人白居易《长恨歌》中的"芙蓉如面柳如眉"，当时文人便将"芙蓉"比喻为面。说法之三是，广东人惯称炒鸡蛋为炒"芙蓉蛋"，于是就把独有的碱水蛋面叫作"芙蓉面"，此为广东人追求的生活美学之体现。云吞面里的面，就连素来并不特别喜爱面食的广州人都能为之折服，足见粤式碱水蛋面的魅力。

The name of "Xi Rong" is from hibiscus, but the "hibiscus" here is used to describe noodles rather than wonton. One of the arguments is that wonton noodles are called "Rong" which is derived from "Furong noodles". "Hibiscus" describes the beautiful face of a woman, so it can be used to describe nice noodles in ancient Chinese literature (the pronunciation of "face" is the same as that of "noodles" in Chinese). The second argument is that "Furong noodles" originated from "willow leaves are like her brows and hibiscus her face" in "Song of Everlasting Sorrow" by Bai Juyi, a poet in the Tang Dynasty. At that time, the literati compared the "hibiscus" to the noodles. The third argument is that Cantonese used to call scrambled eggs as fried "Furong eggs" or "hibiscus eggs", so they called the unique noodles made from flour with alkaline water and eggs "Furong noodles", which

地利 | 第二节 | 小吃不小
Right Place | Section II | More than just snacks

embodies the life aesthetics of the Cantonese. The noodles in the wonton noodles can even impress Guangzhou natives, who are not particularly fond of noodles, showing the charm of the unique Cantonese-style noodles.

面条下碱水是为保证面质爽滑，但碱水味过重，面条味就无立足之地。既要用碱水，又要无碱水味，怎么做？云吞面师傅要做的就是往面团上敷一层纸"过水走碱"。不同季节，面条走碱的时间有差异，在夏天需要用3～4个小时，而在冬天则需要7～8个小时。

The alkaline water is used for the smoothness of the noodles, but if the taste of the alkaline water gets too strong, the flavor of the noodles will be covered. How to get rid of the flavor of the alkaline water? The chef usually lays a piece of paper on the dough to remove the alkaline taste. In different seasons, the time it takes for the dough to go through this process is different, usually three to four hours in summer, while seven to eight hours in winter.

"红花还需绿叶扶"，好面怎么少得了好汤底搭配？云吞面汤底的灵魂是大地鱼，这种鱼的油分充足，煮出来的汤特别浓郁鲜香。有了靓美的大地鱼，这汤底就真的叫"有底"了。其余配料，如虾皮、虾子、猪筒骨、金华火腿、干贝、干蚝等物，均是锦上添花，熬上三四个小时，直至汤色金黄，香味四溢。注意没，这里面可没有绿叶菜哦，通常只会撒上数段韭黄，如此方能保证一碗只有海味香的"黄金汤"。

"Green leaves bring out the shine of red flowers." In the same vein, nice noodles require nice soup. The

■ 老西关云吞面（王维宣摄影）

soul of wonton noodle soup is flounder, which contains abundant fish oil. This feature makes the flavor of the soup particularly rich and delicious. The flounder really lays the foundation for the soup, while shrimp skin, shrimp roe, marrowbones, Jinhua ham, dried scallops, dried oysters and other ingredients are all like the icing on the cake. Bring the soup to boil, then simmer it until the color turns golden and the fragrance is overflowing. Have you noticed? There is no green vegetable but usually a few hotbed chives in the soup, so as to maintain its seafood flavor rather than turn it into a bowl of vegetable soup.

无论是街边小店的掌门人吴锦云、吴锦星兄弟，还是中国大酒店的主厨吴炳亮，都对云吞面有一种近乎变态的苛刻。例如，有些云吞面师傅会把二楼客人的云吞面少煮 5 秒钟，这

地利 | 第二节 | 小吃不小
Right Place | Section II | More than just snacks

5秒是预留给店员将面送上二楼的时间，确保送到客人面前的是足够爽滑弹牙的出品。又例如用竹竿打面，打面机固然不错，高效快速，但吴炳亮却认为，和面时不加一滴水，要全部用鸭蛋来中和，如果鸭蛋黄多了，还要将部分蛋黄捞出，才能保证口感。如此，10斤面粉，竟要用去近5斤鸭蛋；至于压面，则要靠师傅坐在一根硕大的竹子上跳跃加压。吴炳亮选用的竹子都是在从化山里精选出来的5年以上的毛竹，竹节一定要多，这样压得才牢靠。

Be they the brothers Wu Jinyun and Wu Jinxing, the heads of a representative street shop, or Wu Bingliang, the chef of China Hotel in Guangzhou, they are all extremely demanding to the extent of being too harsh on the wonton noodles. For example, some cooks will shorten the cooking time by five seconds for second-floor guests. These five seconds are spared for the waiter to deliver to make sure the wonton noodles are still smooth and al dente. Another example is the use of bamboo to press the dough. The dough pressing machine works well, being efficient and fast. But Wu Bingliang believes that not a single drop of water should be added to the dough, so he uses duck eggs to work as water. In order to ensure the taste, excess duck egg yolks should be taken away. Thus, 5 kilograms of flour should use nearly 2.5 kilograms of duck eggs. As for the pressing part, it relies on the cook to sit and jump on a huge bamboo pole to press the dough. Carefully selected by Wu Bingliang in mountains in Conghua District of Guangzhou, the bamboo has grown for more than 5 years and has many knots, and only with this kind of bamboo can the dough be pressed hard enough.

什么是匠人？这就是匠人。
What is craftsmanship? This is.

对待这样一碗匠人心血烹制的云吞面，必须有正确的姿势：先呷一口汤，合上口，尝其持久鲜味；然后再吃面，一咬即断为爽口，且入口爽滑满蛋香者为上；最后才品味云吞，皮薄滑溜，馅料鲜香。
The right way to enjoy the wonton noodles made of craftsmanship is this: first, take a sip of the soup and savor the long-lasting seafood flavor. Then, eat the noodles. The noodles can be called al dente if one gentle bite should

地利｜第二节｜小吃不小
Right Place | Section II | More than just snacks

separate the noodles, and it's even better if you can taste the flavor of the eggs. Only then can you have the wontons, whose wrappings are thin and smooth, and the fillings delicious.

那么，本节开头的问题，你有答案了吗？没有的话，那就再上一碗云吞面，在品尝中继续寻找答案吧。

So, do you have an answer to the question at the beginning of this section? If your answer is no, why not order one more bowl of wonton noodles and continue to look for the answer?

■ 天鹅酥

广府小饼这一家
Cantonese Pastry

在婚嫁领域，广府人一直保持着"业界良心"，彩礼、礼金、婚宴红包一律都是"讲心不讲金"，这并不意味着广府人不讲究婚嫁礼仪。他们的讲究重点在吃的方面，其中一点就反映在嫁女饼的挑选上。

The Cantonese are most pragmatic in dealing with betrothal gifts and money, who think more of the sincerity than the value. Yet this does not mean that they take no account of the ceremony. Instead, they attach much more importance to eating, which is especially reflected on the choice of the Chinese Wedding Cake.

嫁女饼并非广州人的发明，准确地说，最早是由一个山东人指导一个河北人创造出来的……公元209年，对于河北人刘备来说，是重要的一年，刚刚经历了赤壁之战的辉煌胜利，又从东吴"借"到了荆州这个落脚点，刘备的职业生涯看起来前景一片光明。男人有了事业，爱情也会到来，哪怕是假的。东吴的孙权为了收回被刘备"借"走的荆州，接受了周瑜所献的"美人计"，要来一个"假招亲真绑人质"，试图以自己的妹妹为"诱饵"，把刘备骗到东吴来做人质，以换回荆州。

The year 209 was of significance to Liu Bei, who had just won the Battle of the Red Cliff, and "borrowed" a place called Jingzhou from Sun Quan, Monarch of Dongwu. For a long period Liu didn't return the place to its owner. Therefore, Sun accepted "the stratagem of honey trap" suggested by Zhou Yu, which was to invite Liu to Dongwu to marry sun's sister and capture him as a hostage in exchange of Jingzhou.

那边，山东人诸葛亮掐指一算，识破了东吴人的计策。于是将计就计，授意陪同刘备入赘的士兵一进东吴境内就四处派送礼饼，让刘孙联姻的"喜事"弄假成真。如此一来不仅保住了刘备的性命，还为他带来了一段真姻缘。

Liu's adviser Zhuge Liang was intelligent enough to identify it as a trick. He asked Liu to turn Sun's trick to his own use. Then Liu told his soldiers to deliver cakes to make the message of the marriage public as soon as they arrived in Dongwu, eventually turning the marriage into reality as well as saving his own life.

刀光剑影之间，偏偏有这么一段风流故事。而这风流故事中，广府人又偏偏挑中了几块饼。广府人关注的点是不是有点偏？不管偏不偏，反正这饼广州人是吃定了。

The story is a mixture of fighting and romance, whereas the Cantonese have extracted only one thing from all the complexity——the Chinese Wedding Cake.

按照诸葛亮和刘备的做法，嫁女饼一般由男方在过礼时作为聘礼

地利 | 第二节 | 小吃不小
Right Place | Section II | More than just snacks

■ 广式萨其马

送到女方家中。一方面是为了渲染喜庆的气氛，你想象一下，一帮粗壮大汉抬着一担担礼饼，这一担就是100斤，锣鼓喧天，浩浩荡荡，婚礼之前就已出尽风头。当然，这担数一定要双数，寓意好事成双，俗称"双扁"。至嫁女饼入屋，一一打开，1斤饼分为四头或八头。四头表示1斤4个大饼，八头表示1斤8个小饼。女方家人分装饼食再行派送，关系亲近的亲友一般派2斤左右，较疏远的亲友则派1.5斤左右。要更有面子的，就到莲香楼或陶陶居等知名饼店采购。这质量上乘的嫁女饼，俨然是婚礼前最为尊贵的聘礼。

According to Zhuge and Liu, the Chinese Wedding Cake is sent by the bridegroom to the bride's family as betrothal gifts. What a festive scene when a team of strong men carrying loads of cakes weighing 50 kilograms each swagger through the streets to make the marriage known to the world. Of course, the loads are regularly in doubles in Chinese culture, which signifies that good things come in double. When the cakes are accepted and unpacked, they are divided into fours or eights, that is, four big ones or eight small ones for every 500 grams of cakes. The bride's family will then distribute them to friends and relatives. The closer ones will receive about 1 kilogram of cakes, while the less close, about 750 grams. To reach a higher standard, people usually go to the well-known bakeries like Liangxianglou and Taotaoju to purchase the Chinese Wedding Cake, which is regarded as the most valuable gift of all before marriage.

就这样，女儿的婚讯随着饼香传遍了亲朋好友。

For the bride's family, the message of marrying their daughter spreads along with the cakes.

嫁女饼还有一个美丽的名字：绫酥。绫与衣饰典故有关；而酥者，意为入口酥软、松化。昔日豪门嫁娶以礼饼的丰厚显示体面和气派，绫罗绸缎乃贵族之4款华贵衣料，其中"绫"最名贵。故此，礼饼就以"绫酥"为首选，寓意荣华。绫酥色彩斑斓，有黄、白、红、橙4个颜色：黄绫以豆蓉做馅，寓意贵族和皇气；白绫以爽糖或五仁做馅，代表女方贞洁；红绫最讲究也最贵，以莲蓉做馅，也有用冬蓉，寓意喜庆；橙绫则用豆沙或椰丝做馅，寓意夫妻俩今后生活金灿灿。有些绫酥还加入蛋黄，彰显高贵。很多时候，就是把4种"绫""织"入一个礼盒里，组成"四喜"。

There is another beautiful name for the Chinese Wedding Cake, i.e. Ling Su (Silky Pastry). Ling is a kind of cloth, while Su means crisp and soft In Chinese. The wealthy families in ancient China would choose the most luxurious cakes to distinguish themselves with honor and prestige. Ling is the most luxurious silk among all others, thus the most precious cake is named "Ling Su", which symbolizes glory and wealth. Ling Su has four colors, namely, yellow, white, red and orange. The yellow Ling Su has the green bean paste as its fillings, implying nobility; the white Ling Su has white sugar or mixed nuts as its fillings, identifying the bride with chastity; the red Ling Su is the most expensive and fastidiously made with lotus seed paste or sometimes wax gourd paste, adding a festive atmosphere to the wedding; the orange Ling Su includes bean paste or coconut shreds inside, indicating that the couple will enjoy their golden time after marriage. Sometimes, yolks are added to display gentility. Cakes of the four colors are usually put in the same box to constitute "four kinds of good luck".

　　婚嫁仪式之后，小夫妻的称呼就变成了老公和老婆。饼食竟也跟得上，老婆饼和老公饼横空出世。老婆饼主要源自清朝末年的老字号莲香楼。在莲香楼工作了30多年的黄凤爱说，当时莲香楼有一位潮州籍师傅，师傅对自己制作的点心信心十足，回家探亲就带点心。谁知这位师傅的太太吃了点心之后满脸不屑："你们莲香楼的点心还比不上我娘家炸的冬瓜角呢。"师傅听了妻子的话，很不服气："那就把你娘家的冬瓜角做出来，跟我们莲香楼的点心比一比！"第二天，妻子准备了一锅冬瓜蓉，和上白糖做成馅料，再用面粉皮包成小角，放在油锅里炸至金黄色。师傅品尝后震惊了，于是把妻子做的冬瓜角带回广州给茶楼的师傅们品尝。莲香楼的老板和师傅以为自己什么饼都吃过，也尝了一个。老板说："嗯，味道很好！这是哪里的点心呢，叫什么名字？"潮州师傅一时也回答不

■ 嫁女饼

出来，其中一个师傅便说："这是潮州师傅的老婆做的，就叫它'潮州老婆饼'。"

After the marriage, the couple becomes husband and wife. Accordingly, the Wife Cake and the Husband Cake come into existence.Lianxianglou, founded in the late Qing Dynasty, is famous for making the Wife Cake. The story was told by Huang Feng'ai, who has been working at Lin Heung Tea House for more than 30 years. A pastry chef from Chaozhou was working atLianxianglou. One day, he went back to Chaozhou to see his family with some pastries made by himself, of which he was very proud. However, his wife scorned at his pastries and said, "Your cakes from the prestigious Lianxianglou taste even worse than the wax gourd pastry made by my parents' family." Unwilling to admit defeat, the chef asked his wife to make the pastry mentioned for a comparison. His wife agreed. The next day she filled the lour wrappings with wax gourd paste and white sugar, and put them into a pan and fried them until golden. The chef was shocked after tasting this pastry. He brought it back to Lianxianglou and invited other chefs and boss for a try. Although all of them considered they had tasted the cakes of all varieties around the world, they were tempted by the taste. The boss then asked, "What is the name of this pastry?" One of the chefs replied, "Since this is made by the Chaozhou chef's wife, it can be called 'Chaozhou Wife Cake'."

有老婆饼，自然就会有老公饼。老婆饼正如西关小姐一般小巧细腻，老公饼则个头较大，"面"上还有"麻子"——饼皮上撒满了芝麻。

Where there is a wife, there is a husband. Wife Cake is as delicate as Miss Xiguan from a noble family, while Husband Cake is bigger in size with "pocks" on the surface, which are actually sesames.

旧时西关的大户人家家里少不了婢女，因婢女才有了"鸡仔饼"。这款鸡仔饼可以追溯到清朝咸丰年间，当时的十三行总商伍紫垣同时也是知名粤菜馆成珠楼的主人，他有一个婢女叫小凤。咸丰五年初秋某日，伍紫垣要接待外地客人，碰巧主厨不在。伍紫垣便吩咐婢女小凤做一款广东点心给客人食用，而家里一时无备用食材，小凤便到成珠楼把常年储存的惠州梅菜连同正准备拿来做月饼的五仁饼馅搓碎，加上用糖腌过的肥猪肉，再加上精盐、香料等拌匀，用饼皮包裹，捏成丸形，稍微压扁，放入炉中用慢火烘烤至脆，取出待客。甘香松化、甜中带咸的小饼迅速征服了客人。客人问："此饼何名？"由于此饼出自小凤之手，伍紫垣便随口说是"小凤饼。"

A Xiguan noble family normally had maids, who created another pastry called "Chicken Cake". This pastry comes from a story of a maid during Emperor Xianfeng's reign inthe Qing Dynasty. Wu Ziyuan, Chief of the ThirteenHongs and owner of the famous Cantonese restaurant Chengzhu lou, had a maid named Xiaofeng. One day, Wu's chef was absent when he was to receive a guest from another place. Hence he had to ask Xiaofeng to prepare a Cantonese pastry to serve the guest. Yet there was no foodstuff for pastry at home, so Xiaofeng went to Chengzhulou and brought some preserved vegetable and mixed nuts fillings for mooncakes back, ground and mixed them with sugar-preserved lard, salt and seasoning. She then stuffed the fillings into the dough and flattened it. At last, she baked the pastry in an oven until they turned crisp. The soft, tasty, salty yet sweet pastry won high praises from the guest who was curious about the name of this pastry. Since it was Xiaofeng who made it, Wu casually gave it the name "Xiaofeng Cake".

既然如此美味，具有生意头脑的伍紫垣没有放过这个机会，便让成珠楼的制饼师傅稍做改良，加入南乳、蒜蓉、胡椒粉、五香粉和盐，薄而脆的"成珠小凤饼"正式面世。因小凤饼形状像雏鸡，故又称"鸡仔饼"。鸡仔饼因取名接地气而更受广州人欢迎，一炮而红，香飘粤地，乃至东南亚。

With a great business mind, Wu seized the opportunity at once by asking the chefs in his restaurant to make some improvement by adding fermented red beancurd, minced garlic, pepper powder, five-spice powder and salt. The thin and crisp "Chengzhu Xiaofeng Cake" eventually came into being. The cake was shaped like a chicken, so it was also called "Chicken Cake". This down-to-earth name appealed more to the Guangzhou natives.Then the pastry quickly became popular in Guangdong and even in Southeast Asia.

"嫁女""老婆""老公""小凤",饼食家族,乐也融融。
The Chinese "Wedding Cake", "Wife Cake", "Husband Cake" and "Xiaofeng Cake" consist the Cantonese pastry family , which brings the Cantonese a joyful and happy life.

■ 鸡仔饼

■ 白斩鸡

第三节 | 风味荟萃
Section III | Combining Various Flavours

"有鸡味"
"Chicken Flavor"

有鸡味，是广州人对一只鸡的最高评价。

Keeping the "chicken flavor" is the highest evaluation for the chicken on the dinner table.

为了让一只鸡有鸡味，广州人可以什么都不要——不要酱油、不要麻油、不要酱汁、不要陈醋，更不要辣椒。

To keep this flavor as much as possible, the natives of Guangzhou give up all seasonings such as soy sauce, sesame oil, mixed sauce, mature vinegar and even chili pepper.

白切鸡之名甚妙，"白"者，不加调味品仅用白水清煮而成；切者，斩也，正因不加调料，所以这款鸡可以随时斩开就吃，因而最初的名称是"白斩鸡"。后来大概是人们觉得"斩"字有点不雅，于是改为"切"。

The name "Chopped Plain Chicken" is quite interesting in that "plain" means the chicken is cooked in the boiling water without any seasoning. For this reason, it is convenient to chop the chicken for eating at any time.

白切鸡的做法，清朝已有。乾隆年间的才子兼美食家袁枚写过一本美食随笔《随园食单》，书里的"羽族单"曾经提到过一款"白片鸡"。"片"应是刀切而成，可以合理推测袁大才子提到的"白片鸡"即为如今的"白切鸡"。袁枚是怎么评价"白片鸡"的呢？他写道："鸡功最巨，诸菜赖之，故令羽族之首，而以他禽附之，作羽族单。"大意就是：鸡在宴会中的作用最大，其他菜式都是围绕着鸡展开的。羽族之首为鸡，鸡菜式之首正是"白片鸡"："肥鸡白片，自是太羹元酒之味"。袁枚评价其"尤宜于下乡村、入旅店，烹饪不及之时，最为省便"，还特别提到"煮时不可多"。

The cookery can be traced back to the Qing Dynasty. During Emperor Qian Long's reign, Yuan Mei, an intellect and a gourmet as well, mentioned "Sliced Plain Chicken" in the chapter of "Feathery Family" in his book entitled *A Menu in Suiyuan*. In fact, "sliced" is the synonym of "chopped". So it is reasonable to infer that "Sliced Plain Chicken" is "Chopped Plain Chicken" in modern times. He said, "Chicken plays the most important role in a banquet with all the other dishes beside. Chicken is the head in the "feathery family", while "Sliced Plain Chicken" is the best dish among all chicken dishes. "Chicken in slices has the same flavor as meat soup and liquor." "It is most convenient when there is limited cooking time whether in villages or inns." He especially stressed "not to overcook".

袁大才子不愧为资深美食家，寥寥数笔，便将白切鸡之神韵写全写透了。首先是味道——"太羹元酒之味"，这里的"太羹"是古代祭祀时用的肉汁，而"元酒"就有点意思了，"元酒"即"玄酒"，直接翻译成现代汉语应该是"黑酒"。酒怎么会是黑色的呢？原来，

地利 | 第三节 | 风味荟萃
Right Place | Section III | Combining Various Flavours

中国古人说的"玄酒"虽名为"酒",其实却是水。玄酒是古人在祭祀时当酒用的水。这么说来,所谓的"太羹元酒之味",其实就是鸡肉本身的味,也就是广州人说的"有鸡味"。

An experienced gourmet as Yuan was, he grasped the spirit of this dish with a few words. "Meat soup" was served on the fete ceremony in ancient times, while "liquor" which literally refers to "black liquor" was water in fact, a substitute for liquor on that occasion. In a word, the "flavor of meat soup and liquor" is similar to the taste of chicken itself, which is the "chicken flavor" in the Cantonese saying.

在烹饪条件不足的时候,只要有靓鸡,随便煮开一锅水,就能做白切鸡了。而"煮时不可多",更是点出了白切鸡做法的精髓——不能煮得太熟!想一想,广州人如今吃的白切鸡,也不是全熟的,皮肉之间白里带黄,鸡脂呈啫喱状,此景最是诱人;肉骨之间白里透红,鸡肉与鸡骨交界处透出丝丝殷红,这样的鸡肉才够嫩够滑,而不至于柴。

"Chopped Plain Chicken" is able to be cooked as long as there is chicken and boiling water. Avoiding overcooking is the first consideration. "Chopped Plain Chicken" nowadays is medium well, and it is most delicious when you can see the chicken fat vibrates between the white meat and the yellow skin. The meat is tender enough with blood streak between the meat and the bone.

白里带黄也好,白里透红也罢,奥妙全在于袁枚所说的"煮时不可多"。不过,其实他说的"煮"也未必与如今粤菜师傅的做法相同,更准确地说,现在的粤菜师傅用"浸"。先将鸡宰杀后放入热水中浸泡约

5分钟，取出后快速拔毛并取出鸡内脏。然后放入80℃热水中浸泡，再取出后就要放到冰水里浸泡至冷。像这样冷热交替3回，热水让鸡肉爽滑，冷水让鸡皮爽脆，几出几入之后，最终成就了一只皮脆肉滑的白切鸡。其中，冷热水浸泡时间的掌握至关重要。

The secret of a tasty chicken lies in "not overcooking". However, the cooking in Yuan's time might not be the same as the Cantonese way nowadays. Strictly speaking, the Cantonese chef does not "cook" the chicken, but "soaks" it. The chef soaks the chicken in warm water for five minutes after it is killed, takes it out and quickly plucks the feather, eviscerates it and then soaks it again in water of 80℃, and finally soaks it in ice water until it cools off. By alternating cooling and heating for three times, the chicken becomes tender and the skin crisp and tasty. The most important factor in all is the soaking time.

不过，若有人问，白切鸡怎样才能做得好吃？大部分广州人会告诉你："只有一个秘诀——活鸡，鸡够新鲜，就怎么做都好吃啦。"追求自然的原味，广州人对美味的理解有时就是这么简单。

It is quite interesting when you ask the Guangzhou natives how to make a tasty chicken dish, the majority will tell you that so long as the chicken is fresh, it tastes well however it is cooked. Their understanding of a delicacy is as simple as that freshness makes a perfect dish.

腊味：
来自北回归线的馈赠
Preserved Meat: a Gift from the Tropic of Cancer

广州城北的从化，一座北回归线标志塔矗立其间，这座标志塔，正是热带和北温带的分界线。直射的阳光在这里找到了其北至点，凛冽的北风也在这里找到了其畅行之地。阳光和北风在岭南大地上演了一出"秋日邂逅"，便碰撞出了一种特有的食物——腊味。

The Tropic of Cancer, the line between the tropic and the north temperate zone, goes through Conghua, the northern part of Guangzhou. This line is where the sun is on its northernmost point and where north wind blows in. The encounter of sunlight and north wind in autumn brings about a kind of food that can be found only in this area-the preserved meat.

腊味，显然是岭南平民智慧的产物。在物资匮乏的年代，食物是平民家庭中最宝贵的资源之一，浪费当然是莫大的罪过。然而，没有冰箱，食物如何完好保存呢？于是，人们将肉类收集起来，用盐和酒等作料将其腌制好，便成为腊肉，使得长时间存放成为可能。更有下厨心得的人们，用肠衣来给腌制肉做了一件"外衣"，将肉剁碎，灌入肠衣，便成了腊肠。

Preserved meat originated from the collective wisdom of Lingnan people. In the age of food shortage, food was so valuable to the common people that it was not allowed to waste even a little bit. How to preserve the food without refrigerators became a big challenge. Luckily, the problem was resolved when the meat was soaked and preserved in the sauce made of salt and liquor. This is the origin of preserved meat. Furthermore, some clever people tried to wrap the minces with a casing so that sausages were made.

腊肠和腊肉,给岭南人带来正宗的"腊味",同时也给物资匮乏的岭南人带来终极美味。不过,如今的腊味早已不是穷人家的专利,毕竟,几乎人人都爱腊肠蕴含的那种甘香。

Preserved meat and sausages used to be a treasure to the poor family, but nowadays they are enjoyed more popularly by all people for their flavor.

地利 | 第三节 | 风味荟萃
Right Place | Section III | Combining Various Flavours

早在 1949 年前，专门生产腊味的著名厂家就有皇上皇、八百载和沧洲等。有心人可能会发现问题，腊味不是阳光和北风的"爱情结晶"吗？腊味并不是只有秋风起时才能生产，那其他时间呢？广式腊味的开拓者们当然早就想到了解决方案，那就是"三阵并施"。这是个颇具中国传统文化特色的说法，实际上却包含相当科学的"多种经营"理念：冬季制作腊味，是以火为主的"诛仙阵"；夏季恰恰相反，开设冰室，即以冷为主的"阴风阵"；还有春秋两季的"温和阵"，用制作腊味的下脚料做肥皂。正是这"不务正业"的精神，才让"广式腊味"这块金字招牌得以传承。

Even before 1949, there were several well-known factories which made preserved meat and sausages, such as Huangshanghuang, Babaizai and Cangzhou. Questions had been raised about how to operate the factory throughout the year since preserved meat and sausages were generally made when it turned cold. Professionals had worked out a diversifying operation, that is, to make preserved meat and sausages in winter, to operate an icehouse in summer, and to make soaps with the leftovers of preserved meat and sausages in the mild spring and autumn. It is with this philosophy of diversity that the Cantonese preserved meat and sausages win a leading place in the country.

肉是腊肠的灵魂，业内公认制作腊肠的肥肉和瘦肉的最佳比例应该是 2：8。尽管肥肉是"少数派"，但非常关键，广州的老派美食家一口就能吃出那两成肥肉带来的嚼劲与香气。上好腊肠的肥肉"多一分太腻，少一分太柴"。

■ 经过秋风吹晒，肥仔秋腊肠的味道更为鲜美（马骏、黄紫玮摄影）

The key of making sausages is the meat inside, and the optimum proportion of fat to lean is 2:8. Though a small portion, fat is indispensable. An experienced foodie can easily identify the portion of fat by the chewiness and flavor. The first-class sausage is with the exact proportion of fat, too much of which is greasy while too little, tough.

人靠衣装，腊味也一样，其肠衣是给肉既通风又防腐的"神器"。追求食物原味的广州人最青睐天然猪小肠衣，有了这件密布细微"毛孔"却又薄如纸且富有弹性的"外衣"，腊肠便被赋予了呼吸的能力。当把腊肠放入电饭煲，随着温度升高，肠衣的"毛孔"慢慢张开，吸入清香的"饭气"和水分，肠衣内的五花肉慢慢充盈起来。正是这轻微的扩张，让食客咬下时便觉肉汁丰盈、入口即化。

地利 | 第三节 | 风味荟萃
Right Place | Section III | Combining Various Flavours

Casing is to a sausage what coat is to a person. Casing serves for ventilation and preservation. The best of it is the casing made from pig's small intestine. With this thin, stretchy, multi-pored "coat", a sausage can even breathe. When it is put into the electric cooker to cook, the pores in the casing will open gradually, enabling the sausage to breathe in the rice flavor and water. As a result, the stuff inside the sausage will swell up, which provides a juicy and melting taste for the eaters.

肉是本地新鲜土猪肉,肠衣是猪小肠,晾晒也是天然的北风和阳光,岭南人将食材、气候、光照等种种天然元素汇聚,造就出一种人间美味。

All in all, the Lingnan people make good use of the ingredients, climate and sunshine to create such mouth-watering food.

神奇的是,不同地方的劳动人民同样拥有这种天然的智慧。广州的"邻居"东莞也有自己的品牌腊味——矮仔肠。传说,清末东莞高埗人吕佳个子矮,上街叫卖制作好的腊肠时,常将腊肠拖到地上,其妻建议将腊肠制得短而粗。矮仔肠的制作工艺经过长年发展,在2015年入选广东省非物质文化遗产名录。更有说法,广式腊味其实源于广州的另一个"邻居"中山黄圃。据说黄圃镇上一位卖粥老板,灵机一动,将卖不出去的猪肉、猪肝剁碎,灌入粉肠,便成了最早的腊肠。广州、中山、东莞三市分处粤港澳大湾区东西两岸,却早在100多年前在腊味的甘香上"心有灵犀一点通",岂不妙哉?

It is interesting that people who live outside Guangzhou share the wisdom in making preserved meat and sausages. For instance, the neighbor city Dongguan has established its own brand of sausage called "Shorty Sausage". According to legend, there was a local named Lv Jia in the late Qing Dynasty who made and sold sausages on the street. But he was so short that the sausages he carried always touched the floor. His wife thus suggested that the sausages be made short and thick. After many years' improvement, the production process of "Shorty Sausage" was listed into the provincial intangible cultural heritage list.

吃出那"啫啫"的声音了吗？

Have You Tasted the *Ger Ger* Sound?

看这标题，是不是觉得写错了？其实没错，如果你吃过广州啫啫煲的话。

When seeing the title, do you think it wrong? Actually no, if you have tasted Guangzhou *ger ger* pot.

"哧溜"一声，伙计端着冒着白烟的煲仔豪迈跑过马路，冲到客人面前时，煲盖一掀，立马惹得满桌跃跃欲试。大家早已顾不得绅士、淑女形象，只管举箸争食。原本只是经过、无意消夜的路人，闻到啫啫煲的香味，也忍不住坐下来点菜。这是广州鼎鼎有名的啫啫煲，源于大排档，现纵横粤菜餐桌，在"阳春白雪"和"下里巴人"之间任意转换。

A waiter, carrying a pot in which white smoke is rising into the air, runs across the road in a hurry, rushes to the customers and lifts the lid. Instantly, a big stir comes out from the table. There are no more gentlemen or ladies, but just foodies vying for the delicious dish. Enchanted by the inviting smell of the *ger ger* pot, even a passer-by who is not intended to have night snacks cannot help stopping here and sitting down for an order. This is the famous *ger ger* pot in Guangzhou, originating from *Dai Pai Dong* (a type of open-air food stall) and popular in Cantonese restaurants.

啫啫煲是粤菜独有的烹调菜式，也是煲仔饭的完美延续。"啫啫"实际上是拟声词，瓦煲中的食材经过极高温的烧焗，汤汁不断快速蒸发而发出"嗞嗞"声，"嗞嗞"粤语发音为"啫啫"，于是老广便巧妙地将其命名为"啫啫煲"。啫啫煲成为广州标志性的味道，它所展现的"气"，与小炒的镬气如出一辙。

Ger ger pot is a unique cooking dish in Cantonese cuisine as well as a perfect continuation of clay-pot rice. *Ger ger* is exactly an onomatopoeic word. Through high-temperature roasting and baking, the gravy of those ingredients in a heating clay pot is continuously and rapidly evaporating and sizzling. In Cantonese, the sound of sizzling can be pronounced as *ger ger*. Hence, dishes made by such a cooking method are cleverly named by Guangzhou natives as *ger ger* pots. Food made by the *ger ger* pot does produce the same sensation as that by stir-frying, which makes it a symbolic taste of Guangzhou cuisine.

啫啫煲有生啫与熟啫之分，而最考验厨师功夫的是生啫，当中暗含了广州人对"鲜"的追求。生啫亦与广州独有的烹饪技法"生炒"同源，只不过生炒在镬，生啫在煲。生啫的最大特色在于自然烹调之法，最鲜活的料，一个砂煲，一双筷子，三两下功夫，生鲜的食材在瓦煲里经猛火相焗，由生转熟。这股热腾之气一直连贯到餐桌，趁热吃，最是惹味。

Right Place | Section III | Combining Various Flavours

■ 啫啫煲（王维宣摄影）

There are two cooking methods of *ger ger* pot, one is one-off *ger ger* pot, the other is twice-cooked *ger ger* pot. One-off *ger ger* pot should be made with patience during the sophisticated process, which implies Cantonese's continuous pursuit of freshness. The way to cook one-off *ger ger* pot is similar to stir-frying, a special cooking method invented in Guangzhou. The mere difference between them lies in different tools used-stir-fry demands a wok, while one-off *ger ger* just a pot. The most distinguishing feature of this unique method is its freshness and simplicity, using fresh ingredients with a clay pot, and simply stirring the ingredients with a pair of chopsticks in a short time. Through high-temperature heating, the fresh ingredients gradually become well-done, with steaming-hot vapor constantly emitting into the air even when the pot is served. It would be better to enjoy such palatable food when served hot.

■ 烈火中的啫啫煲 （王维宣摄影）

生啫技法创于广州米其林一星餐厅惠食佳。这家已有27年历史的餐厅，最早开在当时的广州CBD（中央商务区）—— 西关，是他们将啫啫煲做成了城中名菜，并在广州城里掀起一股追随之风。后来餐厅搬到东风东路与滨江西路，进阶版的啫啫煲成为一绝，甚至助力餐厅评上米其林一星餐厅。生啫并非每个厨师都愿意做，因为做生啫要舍得"弃煲"。平均每个煲的使用次数是1.5次。生啫时，煲底常常忍受着"煎熬"，上边是生冷的食材，下边是猛烈大火，一冷一热，做上一两次煲底就裂开了。

The special cooking technique of one-off *ger ger* pot was invented by a Michelin 1-star restaurant in Guangzhou called Wisca Restaurant, a renowned restaurant with 27-year history, which was first set up in Xiguan, the former CBD of Guangzhou. It is Wisca that made *ger ger* pot a well-known Cantonese dish in

地利 | 第三节 | 风味荟萃
Right Place | Section III | Combining Various Flavours

Guangzhou, setting off a special tide of the day with which the whole city started to follow. Later, after the its relocation to East Dongfeng Road and West Binjiang Road, the advanced *ger ger* pot has turned into an excellent delicacy, helping Wisca to be listed as one of star-rated restaurants by Michelin. However, not all the restaurants are brave enough for one-off *ger ger* pot as it costs a lot. As a pot can just be used for 1.5 times on average, the restaurant should change a new one when its life-span is over. When cooking, the bottom of the pot needs to endure the high-temperature and constant heat, suffering from the incitement between cold ingredients above and fierce heat beneath, it will crack and split after being used once or twice.

生于广州的啫啫煲，自然有着与广州人一般的海纳百川之胸襟，包容万般食材。鸡、牛、猪、虾、蟹、贝、蚝、时蔬，甚至木瓜、雪糕都能成为煲中啫物。啫蚝烙把广州的烹调特色与闽南特色菜融为一体，啫木瓜和啫雪糕把两种原本常人以为不可能啫的食材变成受人追捧的美味，这无不彰显出广州人创新改革的精神和力求突破的决心。而"生啫"这种自然烹调法里又蕴含了大道至简的大智慧、高境界，它与白焯、清蒸这些粤式烹饪方法有异曲同工之妙。广州人用这种方式表达自己对自然的敬畏与感恩，感恩自然赐予我们如此优质的食物。

Born in Guangzhou, *ger ger* pot naturally possesses great inclusiveness as the natives of Guangzhou do. It welcomes all ingredients, such as chicken, beef, pork, shrimp, crab, cowrie, oyster and seasonal vegetables,

even papaya and ice cream. For example, *ger ger* oyster omelette is a perfect case of integrating Cantonese and Minnan (southern Fujian) cuisines; while *ger ger* papaya and *ger ger* ice cream turn impossible materials for *ger ger* into amazingly delicious dishes. The innovative spirit of the natives of Guangzhou can be clearly seen in these ingenious inventions. Meanwhile, one-off *ger ger* pot also demonstrates the native's great wisdom and vision of pursuing simplicity. It can achieve the same sensational effects as boiling and steaming. Guangzhou natives use such cooking methods to express their gratitude and respect to nature for endowing them with such excellent food.

在广州，包容万菜

Inclusive Cantonese Dishes Featuring the West and the East

外地友人凡到广州，"五脏庙"毫不寂寞。他们的选择何其多，既可品广州本土的美味，又可以尝到来自全国各地乃至世界各国的风味。当怀念家乡的味道时，说不定也能在此一解乡愁。这就是"食在广州"的包容性与多样性。

If a friend ever comes to Guangzhou, he'll never need to bother what to eat since there are all kinds of food available to him, not only the delicious local food, but also food throughout China or even the whole world. The diversified food here can probably relieve his homesickness for he can taste his homeland food he misses so much in this city. This vividly shows how inclusive and diversified "Eating in Guangzhou" is.

这一特性的历史渊源，可追溯到广州建城时期。2000多年来，广州几乎一直处于对外开放之中，尤其是从明朝开始，更是长期一口通商。即便是明朝海禁严格，规定"片板不许入海"，广州也几乎未被禁过。浙、闽、粤三口岸并存时，"宁波通日本，泉州通琉球，广州通占城、暹罗、西洋诸国"，其余二地亦无广州精彩。

明嘉靖元年撤销了浙、闽市舶司后，广州更是占据一口通商的地位。许多江浙商人将通番商品抵广变卖，再换成广货归浙，这一行为称为"走广"。"食在广州，一定程度上是'走广'出来的。"于是就有了屈大均笔下的"天下食货，粤东尽有"。海外优良之物，多数先从这里登陆，传入广州，再往内陆。清朝，西方国家进入航海时代，广州一口通商，旷世繁华从此开启，"食在广州"的格局渐次形成。

The formation of such features of Guangzhou food can date back to ancient China when Guangzhou was developed to be a city. For more than two thousand years, Guangzhou has been open to the outside world. Particularly since the Ming Dynasty, Guangzhou has served as the exclusive trading port for quite a long period. Even when the Ming Dynasty adopted a strict policy of banning maritime trade with foreign countries and ordered that no boat be allowed to sail on sea, Guangzhou might remain open then. According to historical records, three trading ports once coexisted, that is Ningbo of Zhejiang Province, Quanzhou of Fujian Province and Guangzhou of Guangdong Province. In particular, the trading port in Ningbo offered a passage to Japan, Quanzhou to Ryukyu Islands, while Guangzhou to Champa (Now in Vietnam), Siam (Old name for Thailand) and many western countries, which suggests the superior status of Guangzhou in foreign trade. Moreover, after the Ming government abolished the two customhouses in Zhejiang and Fujian Province in 1522, Guangzhou became an exclusive trading port in China at that time. Many merchants of Jiangsu and Zhejiang Province transported their goods targeted at foreign markets to Guangzhou. After selling these goods to the natives of Guangzhou, they brought goods made in Guangdong back to their homeland. Such a process is referred to as "Trading in Guangdong" which, according to *Taste of Guangdong Food* by Zhou Songfang, to some extent, has contributed to "Eating in Guangzhou" nowadays. It is no wonder that Qu Dajun once wrote, "All food in China can be found in Guangzhou". Back then, most of the quality products from overseas found their way into China through Guangzhou and then were transported to other places. During the Qing Dynasty when the West ushered in a "Navigation era", Guangzhou was the only place in China which had trading relations with foreign countries. This has launched a great prosperity time for Guangzhou and the pattern of "Eating in Guangzhou" also gradually took shape.

通商口岸带来了海外的食物，西餐在国内的最早流行地应当是广州，而非上海。上海早期的西餐馆大多是广东人开的。1885年，广州第一家西餐馆——太平馆开张，当时上海开埠不久。具备创新精神的广州人充分吸收融合西餐的长处，"令广东的茶点及菜肴成中华一绝"。作家程乃珊在《蛋挞与葡挞》一文中指出，这两种食物是充分吸收西点之所长的广东点心代表。"炸牛奶沙拉、奶油焗龙虾、干煎沙碌，都是西菜中做的新美食，莲蓉餐包、香芋餐包，也都是根据西点中的芝士包、奶油包改进的。"

Overseas food spread in China through the trading ports. And it should be noted that it was in Guangzhou

■ 岭南鸡蛋挞（王维宣摄影）

rather than in Shanghai that the western food first went popular. Also, it was Cantonese who opened the early western food restaurants in Shanghai. Guangzhou opened its first western food restaurant—Taipingguan Restaurant in 1885, shortly after Shanghai had opened its commercial port. The Guangzhou natives fully take in the strengths of western food to the local food, which, according to Cheng Naishan, has made Cantonese pastries and cuisine second to none throughout China. Cheng wrote in *Egg Tarts and Portuguese-*

style Egg Tarts that both egg tarts and portuguese are representatives of Cantonese pastries which have fully taken in the strengths of western-style ones. Also, in "The Way I See Guangdong", it is mentioned that fried salad with milk, steamed lobster with cream, dry fried shrimps are all newly-made delicious western cuisines. Steamed buns with sweetened lotus seed paste and fragrant taro stuffing also have their origin from their western counterparts of steamed buns with cheese and cream stuffing.

广州沙面是历经百年的重要商埠,宋、元、明、清时期为中国国内对外通商要津和游览地。侨美食家创始人杨先生生于沙面、长于沙面,他自幼受中西方文化熏陶,是提出"无国界美食"概念并践行的第一人。1960 年,3 岁的他已开始跟随父亲出入顶级西餐厅,比如八重天、大公餐厅、太平馆、经济餐厅。新亚酒店 8 楼知名酒家八重天餐厅建于民国时期,店中有许多中西合璧的菜肴,比如喼汁焗排骨、咕噜肉和松鼠鱼。杨浩益记得,松鼠鱼虽源自西湖得月楼,但它酸甜的风味符合外国人口味,因此广州师傅用舶来调味品李派林喼汁来制作,当时松鼠鱼又叫甜酸鱼块。童年的舌尖烙下深刻的味觉记忆,西餐的基因根植在杨浩益的血液里,这为他日后开创无国界美食打下基础。他主张美味不分国度,所创办的餐厅侨美食家无国界美食沙龙充分融入了全球化理念。

Shamian Island is a centuries-old important trading port in China, a strategically significant crossing for domestic and foreign trade and a popular tourist attraction during the Song, Ming and Qing Dynasties. Mr.Yang, the founder of the restaurant JM Chef, was born in Shamian and grew up there. He was exposed to and influenced by both Chinese and Western cultures when he was a child, being the first to put forward the notion of "no national boundary for delicious food" and made it a guiding principle of his restaurant. In 1960 when he was 3 years old, Yang had begun to accompany his father to dine in top western food restaurants like Bachong tian, Taigong Restaurant, Taipingguan Restaurant, Jingji Restaurant, and so on. On the eighth floor of New Asia Hotel was the well-known "Bachongtian Restaurant", which was built in around 1925. The restaurant offered a number of dishes combining Chinese and Western food ingredients, such as pork ribs steamed in ketchup, sweet and sour pork, squirrel-shaped fish and so on. Yang knew squirrel-shaped fish was originally from

■ 松鼠鱼

Deyue Tower by the West Lake. And since its sweet and sour taste appealed more to foreigners, Cantonese chefs used an imported Lea and Perrins Worcestershire sauce to refine the former version. At that time, squirrel-shaped fish was also called sweet and sour fish lump. The taste of western food has impressed Yang to a great extent and this has laid a foundation for the opening of his own restaurant later. According to him, delicious food belongs to every nation in the world and the vision of globalization is fully embodied in his restaurant.

在这些中西合璧菜中,"喼汁"露面的概率非常高。喼汁是源自英国的调味品,味道酸甜微辣,色泽黑褐。传入广州以后,被广泛运用到粤菜中。另一种舶来调味品咖喱,在粤菜中也充当相当重要的角色。咖喱是最早的全球化产物之一,与粤菜结合用来煮鸡、鸭、鱼蛋、牛肉丸、土豆等。

Worcestershire sauce is frequently employed in the dishes with both Chinese and Western features. Worcestershire sauce, a dark-colored seasoning from Britain, tastes sour, sweet and a little spicy. After it was introduced to Guangzhou, the seasoning was widely used in Cantonese dishes. Curry, another imported seasoning, has also played an important part in making Cantonese dishes. As one of the earliest products of globalization, curry was used in Cantonese cuisine to braise chickens, ducks, fish balls, beef balls, potatoes, and so on.

改革开放以来,广东在生活方式上发生变革,开风气之先,白天鹅宾馆的发展就是证明。20世纪八九十年代,白天鹅宾馆副总经理彭树挺带领行政总厨庄伟佳及团队推动"西菜中做",开创中餐"位上"的新模式,摆盘风格从中式的华丽走向西式的精致。这股风潮,最早是从内地首家中外合资的酒店白天鹅宾馆开始的。"糖醋菊花鱼"正是厨师庄先生在学习欧美国家的摆盘风格之后交出的"成绩单"。这道菜让加拿大宾客惊喜不已,不舍下箸享用。他用芦笋做菊花茎,青瓜做叶,番茄汁等化作"泥土",以鱿鱼或石斑鱼做怒放的菊花。"菊花"形态优雅,色彩搭配柔和。

Since reform and opening up, Guangdong has been a leader in changing lifestyles, which can be presented by the development of White Swan Hotel. In the 1980s and 1990s, under the leadership of the vice manager Peng Shuting, executive chef Mr.Zhuang Weijia and his team sought to "adapt Western cuisine for the Chinese table" and established a new model of "individual servings" in Chinese cuisine. Also, the style of dishing up the cuisine had transitioned

地利 | 第三节 | 风味荟萃
Right Place | Section III | Combining Various Flavours

from being fancy to delicate like its western counterpart. It was in White Swan Hotel, the first Sino-foreign joint venture hotel in China's mainland that such a trend emerged. The cuisine "sweet and sour chrysanthemum fish" was dished up by Zhuang after he had learned the Western style of dishing up the cuisine. This dish impressed Canadian customers so much that they felt it a pity they had to destroy such a beautiful cuisine. Zhuang used asparagus as the stem of chrysanthemum, cucumber as the leaves, tomato juice and ketchup as the earth for chrysanthemum and grass carp or grouper as the blossoming chrysanthemum. As a result, the chrysanthemum looked elegant with a bright color.

尔后，各大五星级酒店、外国餐厅与外地餐厅纷纷涌入广州，带来了多元的食材、烹饪技法以及美学观念，愈加丰富了广州饮食文化，促进了粤菜与外来菜的交流融合，演变出个性独特的新派粤菜。世界三大美食——意大利黑松露、法国鹅肝、法国鱼子酱，接连出现在粤菜餐桌上，连挪威三文鱼、日本黑蒜等健康食材也被粤菜师傅巧妙地运用到菜肴中。为将外国先进分子料理的烹饪方法融入中餐，粤菜师傅广泛地运用慢煮方法来烹饪食材。实际上，这与粤菜中白切鸡的制作方法有异曲同工之妙。粤菜从过去纯粹注意视觉，延伸到嗅觉、味觉、触觉、听觉，粤菜令食客的感官全方位投入。餐盘上，法国精致风与日式自然风相映衬，粤菜呈现极简主义美学与自然美学的摆盘风格。在现代新派粤菜餐厅中，石头、木头、贝壳、海盐、香草、花等自然的馈赠，成为运用广泛的摆盘元素。

Later on, a number of five-star hotels, foreign restaurants and restaurants run by non-locals were gradually opened, bringing diverse food ingredients, cooking techniques and new aesthetic views to Guangzhou. All these have enriched Cantonese diet and promoted exchanges and combining Cantonese dishes with foreign ones. Naturally, a new type of Cantonese cuisine was born. Italian perigord truffle, French foie gras and French caviar—three most delicious dishes in the world have made their way onto the table with Cantonese dishes. Salmons from Norway, black garlic from Japan and other healthy ingredients have been skillfully used by chefs in making Cantonese dishes. The advanced cooking method of molecular gastronomy is applied into cooking Chinese dishes and the chefs are used to simmering Cantonese dishes. In fact, such a cooking technique is quite similar to the one applied to Chopped Plain Chicken. Cantonese dishes have changed from only focusing on the look to improving the smell, taste, touch and hearing. In terms of the style of dishing up the cuisine which incorporates the French elegance and Japanese natural style, Cantonese dishes have developed a style combining simplicity and natural beauty. In modern restaurants with newly-developed Cantonese dishes, stone, wood, shell, sea salt and fragrant herb and other materials from nature have become the widely-employed elements.

广州开放、包容、敢为天下先的个性,在一菜一肴中表露无遗。

The open, inclusive and innovative character of Guangzhou has been clearly demonstrated in each of the Cantonese dishes.

■ 粤式点心精致"装扮"

第三章 | Chapter III

Right People

　　美食，不断聚拢着这个城市的人心。广州的温度，就像茶楼里的那杯茶，温热不烫口；广州的人心，就像丰富的茶点，多彩而又相通。相通、相异又相融的美食文化，不仅给广府人带来了满满的幸福感和归属感，还彰显了广府美食文化的独特魅力。

　　Delicacies have increasingly won the hearts of the people in the city. The temperature in Guangzhou is just like a cup of tea in the teahouse, warm but not scorching; the hearts of the people are just like the varied dim sums, colorful but not isolated. The interconnected, different, and harmonious food culture has brought a great sense of happiness and belonging to the Cantonese; and has further enhanced its unique charm to the outside world.

■ 普洱茶（苏韵桦摄影）

第一节 | 茶楼百态
Section I | Life in the Teahouse

得闲饮茶
Enjoy Drinking Tea at Leisure

"得闲饮茶"是广州人的一句口头禅，也是广州人的一种生活方式。

Duk Haan Yum Cha, which means to enjoy drinking tea at leisure, is an everyday saying of the natives of Guangzhou and also a lifestyle of them.

生活在广州的作家黄爱东西曾这样描述广州人对饮茶的高度热爱："广州人是可以一整天都泡在茶楼里的，早上五点半到十一点喝早茶、吃早餐，十一点到下午两点半吃午饭，两点半到六点喝下午茶，六点到九点吃晚饭，九点到十二点喝夜茶。"

Huang Aidongxi, a writer living in Guangzhou, once described the Guangzhou natives' craze for tea drinking, "They can stay in the teahouse all day long, from 5: 30 to 11: 00 a.m. for morning tea and breakfast; from 11: 00 a.m. to 2: 30 p.m. for lunch; then from 2: 30 to 6: 00 p.m. for afternoon tea; from 6: 00 to 9: 00 p.m. for dinner; and from 9: 00 p.m. to midnight for night tea."

"饮茶"只是一个简称,短短两个字包含了3层意思:喝茶、吃点心和"吹水"。"吹水"即"聊天",一语道破饮茶所蕴含的社交内涵。

Enioy drinking tea is just a phrase made up of three words, but indicates three activities: drinking tea, eating dim sum and chatting, revealing that teadrinking is a way of socializing.

清晨的茶楼是老人家们的主场,他们通常在固定的时间去固定的茶楼,在条件允许的情况下,坐固定的位置。有的老人别看是一人赴茶楼,他们一坐下前后左右都是老友,甚至跟茶楼的服务员都非常熟络,这是日积月累沉淀下来的情感。到了午间,老人家们饮完茶散场,此时茶楼成为家庭聚会或年轻人友聚之地,孩子们的嬉戏声、大人们的闲叙家常、朋友间的高谈阔论融于"一盅两件"之间,这是广府人倍感亲切的生活场景。

The teahouse in the morning is for the elderly. They usually go to the same teahouse at a fixed time and sit on the same seats if possible. Although some elderly people go to the teahouse alone, they are familiar with those who are sitting around in the hall, and even with the waiters. This is the friendship that has been cultivated over time. After the elderly leave at midday, the teahouse becomes a paradise for families or young people. Children's frolic, adults' casual talk and friends' delightful laughter melt into the One Pot and Two Pieces (one pot of tea with two dim sums), making the teahouse a lively place.

过年的茶楼更有氛围,来饮茶的食客通常会备上利市,利市一

人和 | 第一节 | 茶楼百态
Right People | Section I | Life in the Teahouse

■ 普洱茶（苏韵桦摄影）

般都是几块钱一个。食客见到酒楼的服务员会派利市，表示吉祥祝福，而服务员接下祝愿，兴高采烈回敬一句"恭喜发财"。茶楼中的节日气氛就此烘托起来。

The Chinese New Year will add more festivity to the teahouse. Diners who come to enjoy drinking tea will always prepare some *lai see* (lucky money in red envelopes), usually with a few yuan each; then deliver them to every waiter they see, which conveys their good wishes. The waiter will take the lai see and return it with a cheerful wish, *Kung Hei Fat Choy* (wish you prosperity). The festive atmosphere in the teahouse is thus set off.

旧时茶楼的社交属性更强。清咸丰、同治年间，出现为平民大众提供歇脚之地的"二厘馆"，实则是茶楼的前身，是工友歇息解渴之处，这里为平民大众提供的只是粗茶糕点。清康熙二十五年，广州设立粤海关，有了"十三行"。此时商贾名流需要一处款待生意伙伴的场所，第一间现代化茶楼"三元楼"便应运而生。随后，二厘馆被茶楼取代，茶楼数量越来越多。20世纪二三十年代，茶楼高、

中、低档皆有,但只经营茶市,不做正餐。据95岁的粤点泰斗陈勋回忆,那时的茶楼如同一个小社会,聚集着三教九流。谈生意做买卖的、相亲的、朋友聚会的皆有,世间百态、人情世故都凝聚在饮茶之中。广州民俗文化研究所所长饶原生则记得,20世纪80年代末,人人茶楼开创了24小时饮茶风潮,夜茶、音乐茶丰富了人们的夜生活。

The teahouse in the past had even a stronger nature of socialization. During the reign of Xianfeng and Tongzhi in the Qing Dynasty, the Er Li Pavilion, which provided a place for the common people to have a rest, was actually the predecessor of the teahouse. It was a place for workers to relax and quench their thirst with plain tea and snacks. Canton Customs was set up in Guangzhou in 1686. Then the Thirteen Hongs was found. However, merchants and celebrities needed a place to entertain their business partners. Therefore, the first modern teahouse, Sanyuan Building, was built. Later, Er Li pavilions were replaced by more and more teahouses. From the 1920s to the 1930s, there were all kinds of high-end and low-end teahouses, but they only provided tea and did not make meals. Chen Xun, a 95-year-old leading figure in Guangzhou dim sum industry, recalled that at that time the teahouse was like a small society, gathering all sorts of people. There were people doing business, blind dating as well as meeting friends. Everything in the world was all condensed into tea drinking. Rao Yuansheng, director of Guangzhou Folk Culture Research Institute, remembers that in the late 1980s, Renren Teahouse started a 24-hour tea drinking trend. New ways of enjoying drinking tea at night with music have enriched people's nightlife.

正如作家沈宏非所说:"人是社会关系的总和,广东人的社会关系,用滚水一冲,合上盖,全部都汇总在茶壶之中。"茶楼最初只是洽谈生意、交际应酬、礼俗往来的场所,但随着时代的发展,茶楼已演变出更加多元化的生活场景。

As the writer Shen Hongfei said, people are the sum total of all social relations; as for the social relations of Cantonese, it is all fused into a teapot after being washed with boiling water. The teahouse is the initial place for business negotiations, social intercourses and etiquettes. With the development of the times, it has presented more diversified life scenes.

广州人饮茶颇具仪式感。首先斟茶涮杯。将筷子呈45度立于碗中,滚烫的热水沿筷子边滑下,至八分满,涮完筷子再涮茶杯。待一切清洁工作结束后,将水倒于水盅里。用滚水沏一壶靓茶,开始叹茶食点心。与长者一起饮茶,沏茶的任务就是后辈的,

■ 普洱茶（苏韵桦摄影）

这个人要掌控全场的饮茶节奏，不宜斟得太频繁，亦不可吃得太开心而忘了沏茶；茶汤之浓淡冷热，全凭此人细心观察。水滚，泡茶，茶叶似入水的活鱼，上下翻腾，几经浮沉，浮时淡然，沉时坦荡，最后归于平静安居壶底，化作茶汤伴点心。受茶者，手做拳状，手指内扣，轻敲桌面感谢"斟茶"，每每如是。这是饮茶中的"叩谢"礼仪，用导演于文的话说："这是老广的分寸，谈笑风生的爽朗和规矩情理的细腻都不落下。"

The natives of Guangzhou enjoy drinking tea with a sense of ritual: first, they must pour boiling water to rinse the cups. With chopsticks standing at 45 degrees in the bowl, hot water slides down the sides of them until the bowl is almost full. After rinsing all the tableware, pour the water into a special bowl. Then it is time to boil another pot of water to make tea and start enjoying tea and dim sums. When drinking tea with the elderly, it is the young generation's task to make tea. This person should follow the rhythm of drinking tea at table, neither to pour tea too frequently, nor to indulge in eating and forget the tea. The strength the warmth of the tea are determined by this person's careful observation. Make tea when the water boils. Tea leaves are like living fish in the water, rolling and blowing. After several ups and downs, they finally settle down at the bottom of the pot, making the tea agreeable to dim sums. The tea-taker always taps the table half-folded fingers with to thank the one for pouring tea. This is the etiquette of knocking thanks in tea drinking. In the words of the director Yu Wen, this is the discretion of the Guangzhou natives. The hearty laughter and the delicate rules shall both be borne in mind.

就这样，在一抿一品间，广州人开始了一天的生活。
With sips of tea, the natives of Guangzhou start their day in life.

在一抿一品间，广州人结束了一天的生活。
With sips of tea, they end their day in life.

平淡或者精彩，皆是茗香茶点里的烟火人生。
Plain or wonderful, life continues in tasting tea and dim sums.

■ 牛舌酥（王维宣摄影）

一盅两件：粤食"万花筒"

One Pot and Two Pieces: Cantonese Food Kaleidoscope

广州人饮茶，又谓之"叹茶"，一盅两件足以叹世界。所谓"两件"，指点心。最早的"两件"不过是粗糙的松糕、芽菜粉和包点等物美价廉的茶点。后经过点心师创制与丰富，茶楼中的点心花样得以升级，品种得以扩充。

The natives of Guangzhou drink tea as well as enjoying tea, which means to enjoy life with a pot of tea and two pieces of dim sum. The earliest "two pieces" were nothing more than coarse muffins, bean sprouts rice noodles, big buns, and other kinds of inexpensive snacks. Later, more varieties of dim sum with higher quality in the teahouse have been created by pastry chefs.

20世纪20年代末期，广州陆羽居茶楼点心师郭兴首创"星期美点"。这张点心纸上的所有点心种类每星期都有新变化，不与"长期美点"重复。20世纪40年代末，广式点心进入繁盛期。当时各茶楼之间竞争激烈，为了招揽食客，许多上档次的茶楼也争相推出"星期美点"，"星期美点"成为当时餐饮界的潮流。粤点泰斗陈勋当时掌管六国饭店点心部,每周的"星期美点"最少有16款点心,包括8咸8甜;更多时候是20款,拆为12咸8甜或者10甜10咸。相较于"四大天王"这类长期美点,能真正体现一家茶楼的特色与创新水平的,可以说是"星期美点"。

In the late 1920s, Guo Xing, a pastry chef at the Luyu Teahouse in Guangzhou, led the trend of Weekly Special Dim Sums. All dim sums on this menu should be adjusted every week, being different from regular dim sums. In the late 1940s, dim sums entered a period of prosperous development. At that time, there were fierce competitions among various teahouses. In order to attract diners, many high-end teahouses also rushed to introduce weekly special dim sums, which became the trend in the catering industry at that time. Chen Xun, a Cantonese pastry chef, was in charge of the dim sum department of the Grand Hotel des WagonLits at that time. At least 16 kinds of dim sum, 8 salty and 8 sweet, were produced as the weekly special dim sums. More often than not, they produced as many as 20 kinds, 12 salty and 8 sweet or 10 sweet and 10 salty. Compared with regular dim sums such as "Four Heavenly Kings", which were shrimp dumpling, steamed pork dumpling, barbecued-pork-filled bun and egg tart, weekly special dim sums can truly reflect the characteristics and innovation of a teahouse.

推出"星期美点"需要点心师具备丰富的创意和开拓精神。每周总有一天,各大酒楼、茶楼的点心"高手"聚于莲香楼一边饮"夜茶"一边"华山论剑"。各家拿出当周的"星期美点"相互比试,交流研

人和 | 第一节 | 茶楼百态
Right People | Section I | Life in the Teahouse

发经验与市场接受度。"星期美点"是粤点师傅们创造力的结晶，反映了他们求新变革的激情与行动力。当年的许多新品种经过时间印证而最终存留下来，成为今日我们所说的"传统点心"，比如玉液叉烧包，掰成两半时，汁液流淌。

■ "星期美点"点心纸

Making Weekly Special Dim Sums requires pastry chefs to be equipped with strong creative and pioneering spirit. One day in every week, the pastry chefs of famous teahouses or restaurants gathered in Lianzianglou to drink night tea while comparing their dim sums with each other. They all took out the weekly special dim sums to exchange with others and shared their development experience and customer satisfaction. Weekly Special Dim Sums demonstrated the creativity of Cantonese pastry chefs, reflecting their passion for new changes. Many new varieties have survived the tests of the times and eventually become what we call traditional dim sums today, such as "juicy barbecued pork-filled bun", which is juicy and tasty.

20世纪30年代的茶楼只经营茶市不做正餐。陶陶居最早将酒楼与茶楼融合,并赋予了饮茶更加精致的体验。水滚茶靓是饮茶的标准,除了茶靓,水也要靓。陶陶居当时每天会到三元里接白云山九龙泉水,运水的马车声势浩荡入城。及至马车运到,陶陶居小厅门口的红泥火炉便派上用场,以半截榄核做炭,瓦茶煲内沸腾九龙泉水。以此水沏茶。

In the 1930s, the teahouse only served tea without meals. Taotaoju was the first to integrate the restaurant with the teahouse and offered a more exquisite experience of tea-drinking. Boiling water and good tea is a must for drinking tea. Besides tea, water should also be of great quality. Chefs of Taotaoju went to Sanyuanli every day to fetch water from the Jiulong Spring in Baiyun Mountains with great momentum. At the entrance of the small hall was a red clay stove with half an olive nucleus to light the fire; on top of the stove was a pottery pot, and the spring water was boiling in it. This was the water to make tea.

粤点的发展可以分为3个阶段:第一阶段为1949年前,那时候茶楼都会设立星期美点部;第二阶段为1949年至改革开放前,那时原料不够,点心原料主要以红番薯和米为主;第三阶段是改革开放后,原料以及可用酱汁越来越丰富,点心味道更加多样化,出现了许多新点心。泮溪酒家现出品顾问王金镜,当年师从"点心状元"罗坤。1982年,他作为广州市政府派往友好城市日本福冈交流的厨师之一,在接受当地电视台采访时被问及能做出多少种点心时,他答道:"我能做2000种,师傅罗坤可做至少4000种。"这可不是吹的,粤点的皮有30多种,馅料有40多种,排列组合,不断变化,上千种不在话下。20世纪80年代已有4000种,现如今恐怕更不止了,点心堪称粤食中的"万花筒"。

■ 虾饺

Generally speaking, the development of dim sums can be divided into three stages: the first period was before 1949 when teahouses set up a department for weekly special dim sums. The second period was during the tough time from 1949 to 1978. As raw materials were insufficient, dim sums were mainly made of sweet potatoes and rice. Since the third period – the reform and opening up, the raw materials and sauces have become more and more abundant, so the tastes of dim sums have been diversified, and many new varieties have appeared. Wang Jinjing, Panxi Restaurant's production consultant, Wang was taught by Luo Kun, the top pastry chef. In 1982, was one of the chefs sent by the Guangzhou government to Fukuoka in Japan, a friendly city of Guangzhou for cultural exchanges. When asked how many kinds of dim sums he could make in a local TV interview, he replied, "I can make 2000 kinds, and my master Luo Kun can make at least 4000 kinds." This is not exaggerating. There are more than 30 kinds of wrappings and 40 kinds of fillings of dim sums. The different combinations can easily produce thousands of varieties. There have already been 4000 kinds in the 1980s, and probably more now. They can surely be called the kaleidoscope of Cantonese food.

究竟广式点心从何而来？其实广式点心发展至今，可分3类。第一类是"土著原生点心"，即广州本地自古有之并经发展的民间小吃，比如炒米饼、薄脆、荷叶饭、粉粿、端午粽等；第二类是从北方传来，经过本土点心师妙手改善创新而来的点心，比如烧卖、萨其马、灌汤包、包子等；第三类是从海外传入广州，被本土吸收改进的西方糕点，以蛋挞、马拉糕、曲奇为代表。

Many people often wonder about the origin of dim sum. In fact, up to now, dim sum can be divided into three categories. The first category is native dim sum, that is, local folk snacks that have been developed since ancient times, such as fried rice cakes, thin crisp cake, lotus leaf rice, glutinous cake, and *zongzi*. The second category of dim sum comes from northern China and is improved and innovated by local pastry chefs, such as siu mai, sachima, soup dumpling, steamed stuffed bun. The third category is from abroad, absorbed and improved by the local people, such as egg tarts, Cantonese sponge cakes and cookies.

现如今的"两件"，早已不止"两件"，跃于点心纸上的每一件点心，皆是食物与历史交融的缩影。

Today's "two pieces" have long gone beyond two. Each kind of dim sum on the menu is a microcosm of the blending of food and history.

第二节 | 食有亲朋
Section II | Sharing with Families and friends

流光溢彩广府家宴
The Colorful Cantonese Family Feasts

20世纪初，海珠区同德里10号，"太史第"中食风鼎盛，坊间各大酒家唯其马首是瞻。这是有"百粤美食第一人"之称的江孔殷之府邸，因江氏是晚清最后一届科举进士，曾进翰林院，故被称为江太史。太史豪爽喜客，有请无类。太史孙女江献珠曾忆道："贵介王孙，达官显要，中西使节，落难英雄，甚或三山五岳人马，无不以一登太史席上为荣。"江太史家宴赫赫有名，家中食谱有"太史菜谱"之称，名菜"太史蛇羹"正出于此。据说，为保证原材料的供应，江太史甚至在番禺设农场，在萝岗设果林供应府内。

In the early 1920s, the trend of eating in "Taishi Mansion", No.10 Tongdeli in Haizhu District, was at its highest. Every restaurant followed it. The mansion's owner was Jiang Kongyin, also named as "the first person who created Cantonese delicacies". Since Jiang was one of the successful candidates selected from the Qing Dynasty's last highest imperial examinations and had also served in the National Academy, he was also called Court Historian Jiang. The Court Historian was generous and friendly, hosting all people without discrimination. As his granddaughter Jiang Xianzhu has recalled, "Royals, officers, ambassadors, heroes in distress, or even visitors from all over the world were all proud of being invited by him." The family feast of Jiang was so famous that its menu was specially called "the Court Historian's Menu". The well-known "Court Historian's Snake Soup" is exactly from this menu. It is said that to ensure the supply of raw materials for his family feasts, Court Historian Jiang even set up farms and grew fruit-bearing forests in Panyu and Luogang of Guangzhou respectively.

人和｜第二节｜食有亲朋
Right People| Section II |Sharing with Families and Friends

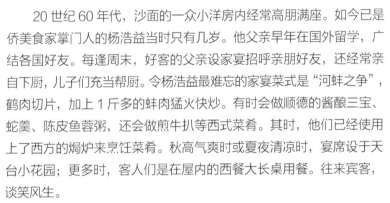

20世纪60年代，沙面的一众小洋房内经常高朋满座。如今已是侨美食家掌门人的杨浩益当时只有几岁。他父亲早年在国外留学，广结各国好友。每逢周末，好客的父亲设家宴招呼亲朋好友，还经常亲自下厨，儿子们充当帮厨。令杨浩益最难忘的家宴菜式是"河蚌之争"，鹤肉切片，加上1斤多的蚌肉猛火快炒。有时会做顺德的酱酿三宝、蛇羹、陈皮鱼蓉粥，还会做煎牛扒等西式菜肴。其时，他们已经使用上了西方的焗炉来烹饪菜肴。秋高气爽时或夏夜清凉时，宴席设于天台小花园；更多时，客人们是在屋内的西餐大长桌用餐。往来宾客，谈笑风生。

In the 1960s, villas of Western styles in Shamian were usually teeming with guests of exalted ranks. Yang Haoyi, the boss of JM Chef Restaurant, was only several years old back then. Yang's father studied abroad in the early days, and made friends with people from around the world. Every weekend, the hospitable father would set family feasts for friends and relatives and cooked dishes himself. The sons would help him. The dish at the feast impressing Yang Haoyi the most was "the battle between cranes and clams". To make this delicacy, sliced meat of cranes and over

500 grams of clams would be stir-fried with oil on high heat. Sometimes, his father would make Shunde's specialties like sauce-brewed aubergines, green peppers and bitter melon, snake soup, fish porridge with citrus peels and also Western dishes like steaks. Even at that time, they had already had the luxury Western oven. On autumn or summer nights with cool winds, feasts were set in a small roof garden; but more often, dinners would be served indoors on a long Western table. Guests came and talked at the table cheerfully.

21世纪的今天，宝华路的西关大屋里，"西关厨神"冼悦华在那备起家宴，招待10多位专程来吃饭的朋友。74岁的华叔精神矍铄，他以老母鸡为原材料，以独特手法处理，做成白切长寿鸡；他以"鲍"会友，自己泡发干鲍、煨制鲍鱼。华叔未从事餐饮业，却是根正苗红的"饮食世家"的传人。华叔的爷爷和父亲曾在香港经营著名的西环八珍酒楼，他耳濡目染亦得真传，退休后闲来无事，拿起旧日家传食谱，做起家宴来。

人和 | 第二节 | 食有亲朋
Right People| Section II | Sharing with Families and Friends

In the 21st century, at one of the Xiguan traditional houses on Baohua Road, "The Xiguan Master Chef" Xian Yuehua is preparing a family feast for 10 friends who come all the way for delicacies. Even at the age of 74, Uncle Hua is still energetic. He uses special methods to make Chopped Plain Chicken with an old hen. He cooks abalone dishes by soaking the dried abalones in water and simmering them. Uncle Hua has never worked in the food industry, but he is an authentic "heir to the catering business". His father and grandfather had operated the famous Bazhen Restaurant in Sai Wan, Hong Kong. Influenced by them, Uncle Hua excels in cooking. Retirement lends Uncle Hua much more leisure time. Since then, he has started to make family feasts according to the menu handed down from his ancestors.

"蓝家菜"的创始人蓝小青生活在广州,他也用家宴的方式记录自家味道。一围蓝家宴往往需要花上3~7天时间做事前准备。"蓝家菜"是他起的名号,以广东菜为基础,博采各地菜系元素后研发自成的一系。其中许多菜肴保留了古代做法,比如铁观音浸墨鱼,是明朝菜。蓝家菜集合了他与他父亲多年的下厨心得。

Lan Xiaoqing, the creator of "Lan Family Cuisine", lives in Guangzhou. He also records his family flavor in the feast. A round of Lan Family feast takes him 3 to 7 days for preparation. The Lan Family Cuisine, named by Lan, has its own characteristics, which, based on Cantonese cuisine, absorbs elements of other cuisines. Many Lan Family dishes preserve the ancient cooking methods. For example, the cuttlefish steeped in Guanyin tea (a variety of oolong tea), is a dish of the Ming Dynasty. The Lan Family Cuisine is cooked according to Lan and his father's years of experience in cooking.

　　"食在广州"的魅力不仅是出街就有好嘢叹，还在于在家有啖靓嘢食。讲究的广州人以四时食材为原料，结合天时与自家口味，在日积月累中形成属于自家的家传菜谱。历代生活在广州这片沃土之上的文人雅士们，更以味道作为纽带，通过家宴的方式传承岭南饮食文化，这是一门艺术，亦体现了生活的智慧。每一道家宴菜的背后都是一个下饭的广府故事。从最早的谭家菜，到今时各式家宴，在流光溢彩的广府家宴中，我们能发现广州人包容开放、热情好客的个性，看见广州人积极向上、豁达乐观的生活态度。

　　The charm of "Eating in Guangzhou" is that one can taste delicacies not only on the street, but also at one's own house. Guangzhou natives have high requirements on their food. They use seasonal raw materials endowed by nature and cook them to their families' tastes. Over time, they have formed family menus unique to them. Moreover, bonded by tastes, generations of literati and scholars living on the fertile soil of Guangzhou carry on the food culture of Lingnan with family feasts. The family feast is an art and contains the wisdom of life. Every dish in a family feast, from the earliest Tan Family Cuisine to today's numerous varieties, tells a Cantonese story that gives people a good appetite. In the colorful family feasts, we can feel the inclusiveness, openness, hospitality, positive spirits and optimism of the Guangzhou natives.

消夜是广州人的仲夏夜之梦

Midnight Snacks:
A Midsummer Night's Dream for Guangzhou Natives

夜影绰绰，荔树婆娑。
花艇艘艘，人声鼎沸。
美味飘飘，食客盈门。

Fascinating night scenes swaying with lychee trees.

Babbles of the crowd from boats with various flowers.

Foodies attracted by smell of food.

荔枝湾畔夜市的盛景，在食物香气和升腾蒸汽中摇曳。特别是在仲夏夜，在没有空调的时代，外出乘凉便成了广州人晚上最重要的休闲方式。逛着逛着就饿了，饿着饿着就上船消夜了。所谓"蛮声喧夜市，海色浸潮台"便是最佳印证，这是唐代诗人张籍写广州夜市的诗句。可见，大唐时的广州已是消夜鼎盛。

On the bank of Lychee Bay, the hustle and bustle of the night market swayed in the smell and soaring steams of food. During the time without air conditioning, enjoying the coolness in the open air was the most important entertainment for the natives of Guangzhou, especially in summer nights. When strolling, people who got hungry would go on board to have something to eat. The lines "babbles of the crowd flood the night market and the boundless sea water soaks the banks" can best prove the lively scene of the night market in Guangzhou. From these lines written by Zhang Ji, a poet in the Tang Dynasty, we can find that night snacks have thrived in Guangzhou then.

休闲时要吃,忙碌后更要吃,勤奋的广州人早就不受限于"朝九晚五"的工作,"炒更"之风一直存在。"更"者,夜间计时单位;"炒"字之玄妙在于"把晚上炒热",意为晚上也保持工作和学习热情。连工作和学习都要用烹饪手法来形容,足见广州人对吃之痴迷。"炒更"后消耗大,需要吃一点夜宵来抚慰劳累了一整天的身心。

Natives of Guangzhou eat something at leisure. They also eat something after work. The diligent people have never been limited by the nine-to-five work schedule, and the trend to "Chaogeng" has long existed. "Geng" is a unit for calculating time at night; "Chao" is an amazing word with the meaning of "heating up", which means to keep the passion of work and study even at night. Natives of Guangzhou use culinary terms even to describe work and study, through which we feel their fascination with eating. Due to the great consumption of energy after chaogeng, they need night snacks to soothe the exhausted mind and body.

■ 艇仔粥

无论是闲人还是忙人，都能在一桌夜宵中找到一天中最后的归宿。

No matter how unoccupied or busy, all people can arrive at their destinations for relaxation with the night snacks.

广州人对美食的追求，小小一条荔枝湾畔显然无法满足。于是到了20世纪20年代，珠江江面成了广州人吃夜宵的新选择。作家欧阳山在《三家巷》里写道："……（张子豪等人）自己轮流出去划桨，小船就在弯弯曲曲的碧绿的水道中……到了宽阔的珠江江面，他们吃过了油爆虾和炒螺片，喝过了烧酒，每人又喝一碗艇仔粥……"

Obviously, it is not enough for the Guangzhou natives to enjoy just the food they have been able to find along the Lychee Bay. So by the 1920s, the surface of the Pearl River had been a new place

for the natives to eat night snacks. In the novel *The Three Family Alley*, Ouyang Shan writes, "…People (including Zhang Zihao) took turns to go out to paddle. And the boat floated on the green curving waterway. After arriving at the wide Pearl River, they ate stir-fried shrimps and fried sliced whelk and drank liquor. And then each ate a bowl of sampan congee…"

从水上又吃到了陆上，转场至骑楼。话说，这骑楼下的消夜更胜一筹，只因夏天此处风凉水冷，此时可上各种夜宵，热气腾腾的猪杂粥、镬气十足的炒牛河，还有一打啤酒……又因冬天遮风避雨，缩进角落，炭火一烧，砂煲一滚，暖意穿梭于骑楼之间。

The place for Guangzhou natives to eat changes from water to land, then to the arcades. Actually, it is better to eat night snacks on the arcades.Because they make guests feel cool in summer, with various stir-fried dishes, hot pig offal porridge, Shahe rice noodles with beef, and a dozen beers; they make guests feel warm in winter, providing shelter for people to huddle up in the corner and feel the warmth passing through the arcades when clay pots are burning over charcoals.

终于，有一天，广州人把最热爱的日常"饮茶"复制了一份到晚上，人称"饮夜茶"。在20世纪40年代末和20世纪80年代皆流行过"夜茶"。雷良是最早一批到广州发展的香港人，他记得1986年流行去大同酒家饮夜茶，或是到人人酒楼喝通宵夜茶。

One day, eventually, Guangzhou natives copied their favorite habit of tea drinking to the night, called drinking night tea. Night tea was popular both at the end of the 1940s and in the 1980s. Lei Liang is one of the earliest Hong Kong people who has a career in Guangzhou. He still has vivid memories that in 1986 people all visited Datong Restaurant at night to drink tea or visited Renren Teahouse to drink tea overnight.

20世纪80—90年代，那是消夜的第一个"黄金时期"。1990年，沿江路长堤的胜记从大排档转为入室经营，从原来10张台扩充到两层共600平方米经营面积。中国香港的一些知名人士皆是座上宾，当时能到胜记吃消夜是身份的象征。以一家之力带旺整条街消夜的传说，正是出自胜记。

The period from the 1980s to the 1990s witnessed the first "golden time" for midnight snacks. In 1990, Seng Kee, located on Yanjiang Road along Changdi (a long dam of the Pearl River), developed from a hawker stall of only ten tables into a two-story restaurant with a total area of 600 square meters. A multitude of celebrities from Hong Kong all visited Seng Kee as a symbol of status. Seng Kee is a legend that has made the whole night snack street prosper.

1991年，流花公园的粥城一开，几十张台全部坐满人，12个厨师都不够人手。掌门人区生大加人手，40个厨师上岗，面积再扩张。他们是全城第一家做粥底火锅的餐厅，一时间200多家餐厅跟风做粥，城中掀起"食夜粥"风潮。当时，南海渔村也做消夜，消夜必点花卷、葱油饼这两款点心。

In 1991, when Liuhua Porridge Restaurant was opened, dozens of tables were full and 12 cooks were not enough. So the boss Mr.Ou, employed 40 cooks and then expanded his restaurant. It was the first restaurant selling porridge hot pots in Guangzhou. Witnessing its success, over 200 restaurants followed to sell porridge, which started an upsurge of "eating night porridge" in Guangzhou. In the meantime, South Sea Fishing Village, a seafood restaurant, started to sell midnight snacks as well and its must-orders are "steamed twisted roll" and "scallion pancake".

1992年，惠食佳在老西关复业，尝试用炭炉啫鸡，后延伸出各式啫煲。"啫啫煲"因此成为消夜爆款，火遍广州城。后来惠食佳不做消夜做正餐，食客也照样追随啫啫煲。

In 1992, Wisca Restaurant had its trial opening in Xiguan. It tried cooking chicken with charcoal stoves and later invented various kinds of *ger ger* pots, which became a must-order midnight snack in Guangzhou. Even though Wisca sells full meals instead of midnight snacks later, many foodies are still crazy about its *ger ger* pots.

人和 | 第二节 | 食有亲朋
Right People| Section II | Sharing with Families and Friends

■ 牛腩牛杂煲（王维宣摄影）

2000年至今，外来风味大放异彩，此刻的广州夜宵兼收并蓄，吸收了外来菜系的特色，如四川风味、东南亚风味、日本风味等，成就独特一派。川式香辣锅、江苏小龙虾……成为羊城夜游者们更多元的选择。活力四射的羊城之夜令无数来访者兴奋，作家易中天在看到广州发达的夜宵市场时就惊叹不已，发出"深夜，才是'食在广州'的高潮……构成独特的广州风景……是地地道道的'广州特色'"之感慨。

Since 2000, as more and more dishes outside Guangzhou have become widely accepted, Guangzhou night snacks have absorbed the features of different styles of dishes, such as spicy Sichuan style, Southeastern Asian style, and Japanese style. There are Sichuan-style spicy hot pots, Jiangsu crawfish...These new night snacks give people in the city more options. The dynamic Guangzhou night excites numerous visitors. For example, the writer Yi Zhongtian, when visiting the booming midnight snacks markets in Guangzhou, marveled, "Night is the climax of 'Eating in Guangzhou'...With night snacks presenting the unique scenery of Guangzhou...showing the real 'Guangzhou feature'."

广州人吃宵夜最任性了，穿着拖鞋、衣服随便一披，小板凳塞在路边就能开吃。细想一下，在宵夜大排档里，不在意白天劳作的疲惫、不回避简陋的环境、不顾忘粗鄙的餐具、不惮于不雅的吃相，还能在暧昧夜色下陪你去吃宵夜直到深夜不归家，可以说是真爱了。一直有说法：与广州人做朋友交心不易，但也许消夜就能给你提供一条与广州人交朋友的捷径，这条捷径不在酒桌上，也不在牌局上，却在你面前那一碗热粥里。

The natives of Guangzhou are most casual when it comes to eating midnight snacks. Dressed casually with slippers on, they sit on small stools on the roadside and start eating. When we think about it, a true friend must be one who is willing to have midnight snacks with you in a snack booth till late at night, despite tiredness from a day's work, the poorly decorated place, the shabby tableware and poor table manners. There is always a saying that it is not easy to be a real friend of the Guangzhou. Midnight snacks, however, may be a shortcut for you to make friends with them. This shortcut is not formal dinners, nor games, but a bowl of steaming porridge in front of you.

毕竟，吃什么不重要，跟谁吃才是关键，对吧？

After all, what matters is with whom you eat, not what you eat.